A2

Do Brilliantly

A2 Chemistry

George Facer

Series Editor: Jayne de Courcy

Published by HarperCollins *Publishers* Ltd
77–85 Fulham Palace Road
London W6 8JB

www.**Collins**Education.com
On-line support for schools and colleges

First published 2002
10 9 8 7 6 5 4 3 2 1

ISBN 0 00 7171773

British Library Cataloguing in Publication Data
A catalogue record for this book is available from the British Library.

Edited by Jane Bryant
Production by Kathryn Botterill
Design by Gecko Ltd
Cover design by Susi Martin-Taylor
Printed and bound by Scotprint

Acknowledgements
The Author and Publishers are grateful to the following for permission to reproduce copyright material:
AQA (pp. 20, 27, 42, 50, 54, 55, 71, 76, 81. 82, 83). Answers to questions taken from past examination papers are entirely the responsibility of the author and have been neither provided nor approved by the AQA.
Edexcel (pp. 12, 16, 22, 29, 30, 37, 44, 45, 51, 64, 74, 79, 80). Answers to questions from past examination papers are entirely the responsibility of the author and have been neither provided not approved by Edexcel. Edexcel accepts no responsibility whatsoever for the accuracy or method of working in the answers given.
OCR (pp. 21, 36, 49, 61, 78). Answers to questions taken from past examination papers are entirely the responsibility of the author and have been neither provided nor approved by the OCR.

Illustrations
Cartoon artwork – Roger Penwill
DTP artwork – Gecko Ltd

You might also like to visit:
www.**fire**and**water**.com
The book lover's website

33784

Contents

How this book will help you
by George Facer

Exam practice – how to answer questions better

This book will help you to improve your A2 Chemistry grade. It contains lots of questions based on the content of the new A2 specifications of AQA, Edexcel, Nuffield and OCR.

The different boards (awarding authorities) have different approaches to the second year of the A level course but this book covers everything except the special topics offered by Nuffield and OCR. All boards set some synoptic questions, and you will find Chapters 10 and 11 especially useful for those.

You should have **two aims** in mind in the final stages of preparation for each of your tests or modules. Firstly, **to maximise your knowledge and understanding of the topics** being tested. Secondly, **to ensure that you have good examination technique** so that you can score as many marks as possible with the knowledge that you have. This book will help you to improve your examination technique.

Chapters 1–9 in this book each deal with a major topic. They are broken down into four separate elements, aimed at giving you as much guidance and practice as possible:

❶ Exam Question, Student's Answer and 'How to score full marks'

The questions at the beginning of each chapter are either taken from specimen papers or are typical of what you will be asked. The student's answers are a mixture of **correct responses and common errors**.

The 'How to score full marks' section explains precisely where the student went wrong. I show you how to pick up those vital extra marks that make all the difference between an ordinary grade and a very good one.

❷ 'Don't make these mistakes'

This section highlights the most common mistakes that students make either in the exam itself or in their preparation for it. When you are into your last minute revision, you can quickly read through all of these sections and make doubly sure that you avoid these mistakes in your exam.

❹ Questions to try, Answers and Examiner's comments

Each chapter ends with a number of exam questions for you to answer. Don't cheat. Sit down and answer the questions as if you were in an exam. Try to put into practice all that you have learnt from the previous sections in the chapter. I've included, before each section, some exam hints which should help you get the correct answers. Check your answers through and then look at the answers given at the back of the book. These are full mark answers.

In the 'Examiner's comments', I highlight anything tricky about the question which may have meant that you did not get the correct answer. By reading through these questions, you can avoid making mistakes in the real exam. Where parts of questions refer to only one or two boards, this fact is clearly marked.

❸ 'Key points to remember'

These pages list some of the most important facts that you need to know, or definitions that you need to learn for each topic. They are not meant to replace your notes, but are a quick check on those points that it is vital to know before going into your exam. You'll find these pages really helpful with your last minute revision.

The major difference between AS and A2 exams is that **some of the A2 questions will be synoptic**. This means that they are designed to bring together ideas, knowledge and skills from all parts of the specification, including topics that you have studied at AS.

Synoptic questions

Synoptic questions are a real test of your ability as a chemist. To succeed you have to revise more than one topic and you must be prepared to link your understanding of one topic to another. Chapter 10 in this book gives you detailed guidance on ways to revise for synoptic questions. Chapter 11 gives examples of the types of synoptic questions set by all the boards, guidance about how to answer them and questions for you to try.

For some topics you may find it useful to look back at *Do Brilliantly – AS Chemistry* if you have it on your shelf.

The topics covered by your specification

The four boards have significant differences both in the order in which the specification is covered and in the content. The chart below lets you see at a glance which chapters in this book refer to the different tests or modules that your board sets.

	AQA	EDEXCEL	NUFFIELD	OCR
Chapter 1 Organic chemistry – reactions	Unit test 4 Unit test 6 A: Synoptic B: Practical skills	Unit tests 4 & 5 Unit test 6 A: Practical skills B: Synoptic	Unit test 4 Unit test 5 A: Practical skills B: Special studies Unit test 6: Synoptic	Unit test 4 Unit test 5/2 to 5/6: Options Unit test 6/2 or 6/3: Practical skills
Chapter 2 Organic synthesis, – analysis and spectroscopy	Unit test 4 Unit test 6 A: Synoptic B: Practical skills	Unit test 5 Unit test 6 A: Practical skills B: Synoptic	Unit test 6 Synoptic	Unit test 4 Unit test 5/2 to 5/6: Options Unit test 6/2 or 6/3: Practical skills
Chapter 3 Chemical equilibrium	Unit test 4 Unit test 6 A: Synoptic B: Practical skills	Unit test 4 Unit test 6 A: Practical skills B: Synoptic	Unit test 4 Unit test 5 A: Practical skills B: Special studies	Unit test 5/2 to 5/6: Options Unit test 6/1: Synoptic, 6/2 or 6/3: Practical skills
Chapter 4 Acid-base equilibria	Unit test 4 Unit test A: Synoptic B: Practical skills	Unit test 4 Unit test 6 A: Practical skills B: Synoptic	Unit test 4 Unit test 5 A: Practical skills B: Special studies	Unit test 5/2 to 5/6: Options Unit test 6/1: Synoptic, 6/2 or 6/3: Practical skills
Chapter 5 Kinetics	Unit test 5 Unit test 6 A: Synoptic B: Practical skills	Unit test 5 Unit test 6 A: Practical skills B: Synoptic	Unit test 4 Unit test 5 A: Practical skills B: Special studies	Unit test 5/2 to 5/6: Options Unit test 6/1: Synoptic, 6/2 or 6/3: Practical skills
Chapter 6 Thermo-dynamics	Unit test 4 Unit test 6 A: Synoptic B: Practical skills	Unit test 4 Unit test 6 A: Practical skills B: Synoptic	Unit test 5 A: Practical skills B: Special studies Unit test 6: Synoptic	Unit test 5/1 to 5/6: Unit test 5/2 to 5/6: Options Unit test 6/2 or 6/3: Practical skills

	AQA	EDEXCEL	NUFFIELD	OCR
Chapter 7 The periodic table – period 3	Unit test 5 Unit test 6 A: Synoptic B: Practical skills	Unit test 4 Unit test 6 A: Practical skills B: Synoptic	Unit test 5 A: Practical skills B: Special studies Unit test 6: Synoptic	Unit test 5/1 Unit test 5/2 to 5/6: Options
Chapter 8 Redox equilibria	Unit test 5 Unit test 6 A: Synoptic B: Practical skills	Unit test 5 Unit test 6 A: Practical skills B: Synoptic	Unit test 5 A: Practical skills B: Special studies Unit test 6: Synoptic	Unit test 5/1 Unit test 5/2 to 5/6: Options Unit test 6/2 or 6/3: Practical skills
Chapter 9 Transition metals	Unit test 5 Unit test 6 A: Synoptic B: Practical skills	Unit test 5 Unit test 6 A: Practical skills B: Synoptic	Unit test 5 A: Practical skills B: Special studies Unit test 6: Synoptic	Unit test 5/1 Unit test 5/2 to 5/6: Options Unit test 6/2 or 6/3: Practical skills
Chapter 10 How to revise for synoptic questions	Unit test 6: Synoptic on all AS and A2 content	Unit test 6: Synoptic on all AS and A2 content	Unit test 6: Synoptic	Unit test 6: Synoptic
Chapter 11 Synoptic questions	Unit test 6: Synoptic on all AS and A2 content	Unit test 6: Synoptic on all AS and A2 content	Unit test 6: Synoptic	Unit test 6: Synoptic

Exam tips

- **Know the rubric**. Do you have to answer all the questions? How much time do you have for each question? As a rough guide the rate is about 1 mark per minute.

- If a question is broken down into sections, **consider each section as part of the whole**, because the various parts will be linked.

- Read each section carefully before answering it.

- The total marks for each part will be shown in or near the margin. **If there are two marks for a part, make sure that you write at least two statements that are worthy of scoring one point each.**

- **Set out calculations clearly**. Give your final answer to the correct number of significant figures (if in doubt give it to 3) and add the unit. Don't round up intermediate answers to 1 or 2 significant figures.

- **If you want to alter something**, neatly cross it out and write your answer in a free space in the question paper. If it is on a different page, alert the examiner by writing 'see page xx'. If you

want the examiner to mark what you originally wrote, write 'please ignore crossing out'.

- If you are asked for a name, **don't give alternatives**. If one is wrong, you will be penalised.

- When you have finished, **check as many of your answers as you have time for**:
  - **Start with a calculation**. Have you worked out the relative molecular mass correctly? Have you misread any numbers? Re-calculate the answer to check that you have not made a calculator error. Finally check the significant figures and units.
  - **Check all chemical formulae**. Make sure that you have not made any silly errors. Check the oxidation state in names.
  - **Check that all equations balance**.
  - **Beware of contradictory statements** such as conc H_2SO_4(aq) or NaOH in acid.
  - Check that you have given **the full name or formula** for a reagent.
  - In organic chemistry **check formulae** - do all atoms have the right number of bonds?

1 Organic chemistry - reactions

Exam Questions and Student's Answers

1 Compound **A** has the formula $H_2C=CH-CH_2-CHO$.

(a) State what would be observed when compound **A** reacts with Fehling's solution. Give the structure of the organic product and state the type of reaction occurring,

Observation: The solution goes red ✗

Structure of product: $H_2C=CH-CH_2-COOH$ ✓ Type of reaction: Oxidation ✓ [3 marks] 2/3

(b) **A** also reacts with bromine. What would you observe in this reaction? Give the structure of the product and state the type of reaction occurring.

Observation: The bromine goes clear ✗

Structure of product: $H_2BrC-CHBr-CH_2-CHO$ ✓ Type of reaction: addition [3 marks] 1/3

(c) Using RCHO to represent compound **A**, write the equation for its reaction with hydrogen cyanide and give the mechanism for this reaction.

Equation: $RCHO + HCN \rightarrow RCH(OH)CN$ ✓

Mechanism:

 [4 marks] 2/4

(d) The product in (c) is a mixture of two isomers.

(i) Draw the structures of the two isomers, showing clearly the way in which they differ.

✗✓

[2 marks] 1/2

(ii) Name the type of isomerism and point out what in the structure causes it.

It is optical isomerism. ✓ The central carbon atom is chiral, so the isomers are non-superimposable mirror images of each other. ✓

[2 marks] 2/2

(iii) Explain what, if anything, would happen if plane polarised light were shone through the solution obtained in (c).

The light would be bent ✗ one way by one isomer and the other way by the other isomer. ✗

[2 marks] 0/2

8/16 [Total 16 marks]

7

How to score full marks

1 (a) The correct observation is that 'a red precipitate is obtained'. You must ensure that you **include both the colour and the physical state** (bubbles of gas/precipitate, etc.) when an observation is asked for.

(b) The correct observation is 'the brown bromine goes colourless'. **Clear does not mean the same as colourless.** For a question on the type of reaction, you must include the words electrophilic, nucleophilic or free radical as well as addition, substitution or elimination. Here, it is 'an electrophilic addition reaction'.

(c) The student made two common errors. The first is that the **curly arrow must start from the correct atom**, which is the carbon atom in the CN⁻ ion, not the nitrogen atom. The second mark was awarded for a correctly drawn curly arrow from the π bond in the C=O to the oxygen atom, but a mark was then lost for failing to put the resulting negative charge on the oxygen atom. The correct mechanism is:

(d) (i) **You must draw a three-dimensional** representation when drawing **optical isomers**, and you must draw them as **mirror images** of each other. The correct answer is shown on the right.

(iii) The correct answer is 'there will be no effect on the plane polarised light, as the racemic mixture of the two isomers is produced in the reaction'.

Don't make these mistakes...

If you are asked to give a **full** (displayed) **structural formula**, you must show **all the atoms separately** and include **all the bonds**. Hydrogen atoms must not be omitted.

Formulae must not be ambiguous (e.g. $C_2H_4Br_2$ could be CH_2BrCH_2Br or $CHBr_2CH_3$).

If there are four different groups on a carbon atom, it will show **optical isomerism**, but don't say that it will bend light: it **rotates the plane of polarisation**.

Conditions should be written over the arrow or listed separately.

In mechanisms, the **curly arrow** must start either **from a bond** or from a **lone pair of electrons**. It must go **towards** an atom forming either a **negative ion** or a **covalent bond**.

Key points to remember

Isomerism

- **Geometric** (*cis/trans*) occurs when there is a **C=C group** in a molecule, and the carbons have two different groups attached to them (an **asymmetrical** alkene). For example, there are two geometric isomers of but-2-ene:

$$H_3C-CH_3 \quad H_3C-H$$
$$\underset{\text{cis} -}{C=C} \quad \underset{\text{trans} -}{C=C}$$
$$H \quad H \qquad H \quad CH_3$$

- **Optical** occurs when there is a **chiral centre** in the molecule, such as a carbon atom with four different atoms or groups attached to it. For example, there are two optical isomers of 2-hydroxypropanoic acid (lactic acid):

$$\underset{H_3C}{\overset{COOH}{C}}\text{---OH} \qquad \underset{HO}{\overset{HOOC}{C}}\text{---}CH_3$$
$$H \qquad\qquad H$$

One isomer will **rotate** the plane of polarisation of plane polarised light **clockwise** and the other anticlockwise.

Equimolar amounts of the two optical isomers result in a **racemic mixture**, which has **no effect** on plane polarised light.

Aldehydes

- These contain the **CHO group** (e.g. ethanal CH_3CHO).
- **Preparation**: add a mixture of **dilute sulphuric acid** and **potassium dichromate** solution to a **hot primary alcohol**. The aldehyde distils off.

- **Reactions**: nucleophilic addition and oxidation.

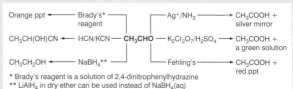

* Brady's reagent is a solution of 2,4-dinitrophenylhydrazine
** $LiAlH_4$ in dry ether can be used instead of $NaBH_4(aq)$

Ketones

- These contain the ⬡ group (e.g. propanone, CH_3COCH_3).
- **Preparation**: heat a **secondary alcohol** under **reflux** with a solution of **potassium dichromate** and **sulphuric acid**.

- **Reactions**: ketones react by nucleophilic addition, but are not oxidised.

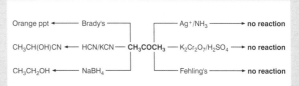

Benzene, ⬡, C_6H_6

- Reacts by **electrophilic substitution**.

	conc. $HNO_3H_2SO_4$ at 60°C	$Br_2(l)$ Fe or $FeBr_3$ catalyst	
$C_6H_5NO_2 + H_2O$ ←			→ $C_6H_5Br + HBr$
$C_6H_5C_2H_5 + HCl$ ←	C_2H_5Cl dry $AlCl_3$ catalyst	CH_3COCl dry $AlCl_3$ catalyst	→ $C_6H_5COCH_3 + HCl$

Reactions of phenol ⬡ C_6H_5OH (not AQA)

Reacts with	Conditions/observations	Product
bromine	aqueous – no catalyst; white precipitate formed	2,4,6-tribromophenol + HBr
sodium hydroxide solution	liquids become miscible	sodium phenoxide, $C_6H_5O^-Na^+$

Reactions of amines

- These contain the NH_2 group e.g. ethylamine, $C_2H_5NH_2$

Reacts with	Conditions	Product
acids, e.g. H^+ ions	room temperature	$C_2H_5NH_3^+$
halogenoalkanes, e.g. C_2H_5Cl	heat in alcoholic solution	$(C_2H_5)_2NH$ + HCl
acid chlorides, e.g. CH_3COCl	mix at room temperature	$CH_3CONHC_2H_5$ + HCl

Preparation of amines

- **Substitution** of halogenoalkanes with ammonia. Heat chloroethane with excess ammonia in alcoholic solution in a sealed tube:

$$C_2H_5Cl + 2NH_3 \rightarrow C_2H_5NH_2 + NH_4Cl$$

- **Reduction** of nitriles (cyanides) with $LiAlH_4$:

$$CH_3CN + 4[H] \rightarrow CH_3CH_2NH_2$$

- **Reduction** of nitrobenzene with tin and concentrated HCl:

$$C_6H_5NO_2 + 6[H] \rightarrow C_6H_5NH_2 + 2H_2O$$

Reactions of amino acids

- e.g. aminoethanoic acid (glycine), NH_2CH_2COOH

$$NH_3^+CH_2COOH \xleftarrow{\;H^+(aq)\;} NH_2CH_2COOH \xrightarrow{\;OH^-(aq)\;} NH_2CH_2COO^- + H_2O$$

Carboxylic acids

- These contain the $-C\begin{smallmatrix}O\\O-H\end{smallmatrix}$ group, e.g. ethanoic acid, CH_3COOH

- They are weak acids and so ionise in aqueous solutions: $CH_3COOH(aq) \rightleftharpoons H^+(aq) + CH_3COO^-(aq)$

Reacts with	Observation/conditions	Equation
bases	aqueous solution	$CH_3COOH + OH^- \rightarrow CH_3COO^- + H_2O$
carbonates	fizzes, liberating CO_2	$2CH_3COOH + Na_2CO_3 \rightarrow 2CH_3COONa + CO_2 + H_2O$
alcohols	reversibly with conc. H_2SO_4	$CH_3COOH + C_2H_5OH \rightleftharpoons CH_3COOC_2H_5 + H_2O$ (an ester)
$LiAlH_4$	in dry ether solution	$CH_3COOH + 4[H] \rightarrow CH_3CH_2OH + H_2O$
phosphorus(V) chloride	steamy fumes of hydrogen chloride	$CH_3COOH + PCl_5 \rightarrow CH_3COCl + POCl_3 + HCl$

Esters

- e.g. ethyl ethanoate, $CH_3COOC_2H_5$.

Reacts with	Conditions	Equation
alkali	heat under reflux; complete hydrolysis	$CH_3COOC_2H_5 + NaOH \rightarrow CH_3COONa + C_2H_5OH$
acids	heat under reflux; reversible hydrolysis	$CH_3COOC_2H_5 + H_2O \rightleftharpoons CH_3COOH + C_2H_5OH$

Acid chlorides

- Contain the $-C\begin{smallmatrix}O\\Cl\end{smallmatrix}$ group, e.g. ethanoyl chloride, CH_3COCl

$CH_3COOH + HCl \longleftarrow H_2O$ $NH_3 \longrightarrow CH_3CONH_2 + HCl$

$CH_3COOC_2H_5 + HCl \longleftarrow C_2H_5OH$ CH_3COCl

$CH_3COOC_6H_5 + HCl \longleftarrow C_6H_5OH$ $C_2H_5NH_2 \longrightarrow CH_3CONHC_2H_5 + HCl$

Grignard reagents (Edexcel only)

- e.g. C_2H_5MgBr
- **Preparation:** Mix a halogenoalkane and magnesium in dry ether solvent under reflux, with a trace of iodine as catalyst.

$$C_2H_5Br + Mg \rightarrow C_2H_5MgBr$$

- **Reactions:** contain a δ^- carbon atom, which acts as a nucleophile, and attack δ^+ carbon atoms in aldehydes, ketones and in carbon dioxide. The initial product has to be hydrolysed with water or dilute acid.

secondary alcohol ⟵ aldehydes ————————— ketones ⟶ tertiary alcohol

C_2H_5MgBr

carboxylic acid ⟵ carbon dioxide ————————— water ⟶ an alkane

Mechanisms

- **Electrophilic addition (alkenes)** e.g. addition of bromine to ethene

Step 1. The π electrons in the C=C bond move towards one of the Br atoms and polarise the σ bond in Br_2, making one Br atom $\delta+$. A Br^+ ion then adds on.

Step 2. A pair of electrons on the Br^- forms a bond with the + carbon ion

(Also the addition of HBr, when $H^{\delta+}$ is the electrophile, and H^+ adds on).

- **Electrophilic substitution (benzene)** e.g. the nitration of benzene

Step 1. The sulphuric acid reacts with the nitric acid to form the electrophile NO_2^+

$$2H_2SO_4 + HNO_3 \rightarrow NO_2^+ + H_3O^+ + 2HSO_4^-$$

Step 2. The π electrons in the benzene ring bond with the NO_2^+ ion

Step 3. The intermediate loses a H^+ to HSO_4^-, reforming sulphuric acid

(Also bromination (electrophile is Br^+), and Friedel-Craft (electrophile $C_2H_5^+$ or $CH_3C^+=O$))

- **Nucleophilic addition (aldehydes and ketones)** e.g. addition of HCN.

Step 1. Some base must be present, as the first step is the attack by CN^- ions made from base + HCN.

Step 2. The O^- removes a H^+ from a HCN molecule, reforming CN^-

- **Nucleophilic substitution (halogenoalkanes)** e.g. the hydrolysis of iodoethane. A pair of electrons forms a bond with the $\delta+$ carbon atom, and, at the same time, the C—I bond breaks

(The mechanism of the reaction with ammonia is similar, with the lone pair of electrons on the N in NH_3 as the nucleophile.)

Questions to try

Examiner's hints *for question 1*

(a) You may choose nitration, halogenation or the Friedel–Craft reaction. Before you decide, check that you can answer part **(d)** of the question. You might find it easier to discuss the reactions of a ketone and hence choose Friedel–Craft in **(a)**.

(c) Do not give standard safety precautions such as wearing safety glasses.

(d) Your answer must relate to the functional group that you added in **(a)**, not the benzene ring.

(e) The question is not well worded. It means that the enthalpy change is –120 kJ for each C=C. Don't forget to draw an enthalpy level diagram showing the relative energies of 'cyclohexatriene', benzene and cyclohexene.

Q1

The characteristic reaction of benzene is **electrophilic substitution**.

(a) Select a reaction of benzene which illustrates this type of reaction. Give the reagents, the equation, the conditions under which it occurs and the name of the organic product for the reaction you have chosen.

....

....

....

....

[4 marks]

(b) For the reaction selected in **(a)**:

(i) identify the electrophile; ....

(ii) give an equation to show its formation; ....

(iii) give the mechanism for the substitution reaction

[6 marks]

(c) Give two **specific** safety precautions you would need to take in **carrying out** the reaction in (a).

....

....

[2 marks]

(d) Give one reaction of the functional group which was introduced into the benzene ring as a result of the reaction performed in (a). Give the reagents, conditions and equation for this reaction.

....

....

[3 marks]

(e) The enthalpy of hydrogenation of a single $C{=}C$ bond is of the order of -120 kJ mol^{-1}.

(i) Assuming that benzene consists of a ring with three separate double bonds, predict the enthalpy change for the reaction:

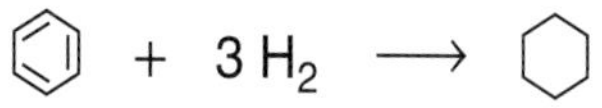

$$\text{⬡} \;+\; 3\,H_2 \;\longrightarrow\; \text{⬡}$$

....

(ii) The enthalpy of hydrogenation of benzene is actually -205 kJ mol^{-1}. What can you deduce from this and your answer to part (i) about the stability of the benzene ring? Use an enthalpy level diagram to illustrate your answer.

[3 marks]

[Total 18 marks]

Examiner's hints *for question 2*
(a) Make sure that you draw the molecule with the correct bond angles and
    that you answer all three parts
(b) What type of bonding makes a substance water soluble?
(c) Which of the two functional groups will react with an alcohol?
(d) Which of the two functional groups will be reduced by LiAlH$_4$?

Q2

If there are two functional groups in a molecule, they normally behave independently of each other. But-2-enoic acid, $CH_3CH=CHCOOH$, illustrates this.

(a) Draw the two stereoisomers of but-2-enoic acid, name the type of isomerism and explain why they are different compounds.

[4 marks]

(b) The acid is insoluble in water, but forms an aqueous solution when dilute sodium hydroxide is added. Explain.

....

....[2 marks]

(c) When but-2-enoic acid is warmed with ethanol and a few drops of concentrated sulphuric acid, a volatile fruit smelling liquid is produced.

(i) Write the equation for this reaction

....[1 mark]

(ii) What is the function of the sulphuric acid?

....[1 mark]

(d) Write the structural formula of the organic product obtained when but-2-enoic acid is reacted with lithium aluminium hydride, LiAlH$_4$, in dry ether, followed by hydrolysis with aqueous hydrochloric acid.

....

....[2 marks]

The answers to these questions are on pages 84–85.   [Total 10 marks]

2 Organic synthesis, analysis and spectroscopy

Before you read this chapter, make sure that you have revised Chapter 1 in this book and Chapter 9 in *Do Brilliantly – AS Chemistry,* if you have a copy. The questions in this chapter require a full knowledge of all the organic chemistry in AS and in A2.

Exam Questions and Student's Answers

1 Propan-2-ol, $CH_3CH(OH)CH_3$, is manufactured from oil and is a useful starting material for the preparation of a number of substances.

(a) Predict the number of peaks and their relative intensities that are observed in the NMR spectrum of propan-2-ol. Explain your answer.

It will have 4 peaks ✗ because the chemical shift, δ, ✓ is different for the H in CH than the H in OH which is different from the H in CH_3. ✓ The intensities are in the ratio 3:1:1:3 because that is the ratio of the different hydrogen atoms. ✗

[4 marks] 2/4

(b) Its mass spectrum shows peaks at *m/e* values of 15 and 45. Give the formulae of the species that cause these peaks.

CH_3, and $CH_3CH(OH)$ ✗ ✓

[2 marks] 1/2

(c) Outline, giving the names or formulae of the reactants, how propan-2-ol can be converted, in at least three steps, into 2-hydroxy-2-methylpropanoic acid, $(CH_3)_2C(OH)COOH$.

$$CH_3CH(OH)CH_3 \xrightarrow{[O]} CH_3COCH_3 \;✓ \xrightarrow{HCN\;✓} CH_3C(OH)(CN)CH_3 \;✓$$

then

$$CH_3C(OH)(CN)CH_3 \xrightarrow{OH^-(aq)\;✗} (CH_3)_2C(OH)COOH$$

[5 marks] 3/5

(d) Explain whether 2-hydroxy-2-methylpropanoic acid is chiral or not.

It is not chiral

[1 mark] 0/1

[Total 12 marks] 7/12

2 The structures of the two isomers, X and Y, are:

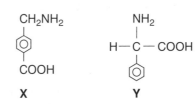

X   Y

(a) Consider **three** different spectroscopic methods which you could use in an attempt to distinguish these isomers. Indicate the results you would expect, supporting your answer with data where appropriate.

Mass spectra: X will have a peak at 30 due to $NH_2CH_2^+$, ✓ but Y will have a peak at 74 due to the $NH_4C_2O_2^+$. ✗

Infra-red: both will have peaks at 1720 cm^{-1} due to C=O and at about 3200 cm^{-1} due to the NH_2 group, ✓ but the fingerprint regions at lower frequencies will be different. ✓

NMR: both will have peaks due to the NH_2 and COOH hydrogen atoms, but X will have two other peaks with heights in the ratio of 2:4 ✓ caused by the two CH_2 hydrogen atoms and the four CH hydrogen atoms, whereas Y will have only 1 other peak due to the CH hydrogen atoms. ✗

[6 marks] 4/6

(b) Only one of the isomers is chiral. State the physical property caused by the chirality which would allow you to distinguish the isomers.

It will rotate the plane of polarisation of plane polarised light. ✓

[2 marks] 1/2

(c) Isomer **X** can be converted to compound **Z**.

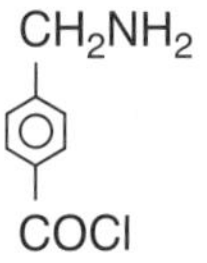

CH₂NH₂

Z

COCl

Molecules of compound Z can react to form a polymer. Draw a section of the polymer, showing at least two monomer units, and indicate any other products which form.

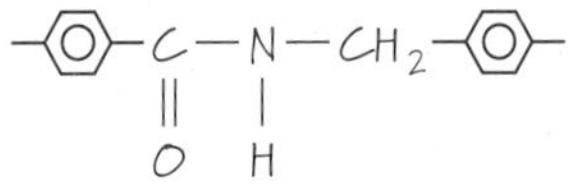

$$\text{--O--C--N--CH}_2\text{--O--}$$
$$\quad\quad \| \quad |$$
$$\quad\quad O \quad H$$

^   ^

[2 marks] 0/2

5/10

[Total 10 marks]

How to score full marks

1 (a) The student did not understand that the two methyl groups have the **same environment** and so will have the **same chemical shift**. The correct answer is: 'There will be 3 sets of peaks. One is due to the two CH_3 groups, which have the same chemical environment, and hence the same chemical shift. Another is due to the hydrogen in the CH group and the third is caused by the OH group. The relative intensities of the three will be in the ratio of 6:1:1 as that is the ratio of the number of hydrogen atoms in each chemical environment.'

(Note that, with high-resolution NMR spectra, the CH_3 peak will be a doublet due to the splitting caused by the single H on the neighbouring carbon. The CH peak will be split into 7 by the six hydrogen atoms on the adjacent CH_3 groups. The OH peak will not be split because it is hydrogen bonded.)

(b) The student made the common error of not writing the charges on the ions. **Only positive ions will be detected in a mass spectrometer.** The correct answer is: '$(CH_3)^+$ and $(CH_3CHOH)^+$.'

(c) The student realised that, as the carbon chain had to be lengthened by one, a cyanide would be a sensible reagent in the scheme. HCN adding to a carbonyl will do this, and so the first step is oxidation of the secondary alcohol to a ketone. However, the student failed to **name or give the formula of the oxidising agent.** [O] can be used in an equation, but the reagent had to be identified. It is 'potassium dichromate(VI) + dilute sulphuric acid.' Another mark was lost for giving OH^- as the hydrolysing agent. This will produce the sodium salt, rather than the acid itself. The best reagent is 'dilute sulphuric acid'.

(d) The question asked for an **explanation**. The reason that the product is not chiral is that 'there are **not** four different groups on the central carbon atom and so the product will be identical to its mirror image'.

2 (a) **Mass spectrum**: the student spoiled a good answer by giving an added up formula for the ion with a m/z of 74. This is a common error. The correct answer is: 'X will have a peak at 30 due to $(NH_2CH_2)^+$, but Y will have a peak at 74 due to the $(NH_2CHCOOH)^+$ ion.'

NMR spectrum: The student gained the first mark for a correct statement about X's spectrum. However, Y will give two peaks due to the CH hydrogen atoms, because they are in different environments. The second mark would be gained by the response: 'Y will give one peak due to the aliphatic CH hydrogen atom, and another, with a height five times bigger, caused by the H atoms on the benzene ring.'

(b) **The student did not say which isomer is chiral.** The correct answer is: 'Y is chiral and so it will rotate the plane of plane polarised light, whereas X will not.'

(c) The student made two mistakes; the structure is wrong and no mention was made of the other product. **You must make sure that you answer a question fully.** The following would score full marks.

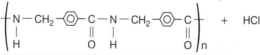

Don't make these mistakes...

Don't forget that **PCl₅ gives steamy fumes** with both alcohols and with carboxylic acids.

Even if a product has a chiral centre, the resulting solution **will not affect polarised light**, as the racemic mixture will be produced.

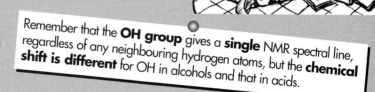

Don't forget that **hydrogen bonding causes a broadening** of the infra-red OH absorption band in alcohols and acids.

Remember that the **OH group** gives a **single** NMR spectral line, regardless of any neighbouring hydrogen atoms, but the **chemical shift is different** for OH in alcohols and that in acids.

Don't forget to **use the spectral data provided** in the question or in a data booklet.

Alcohols do not react with cyanides but halogenoalkanes do substitute with CN^-.

Key points to remember

Synthesis

There are several ways to increase the carbon chain length.

- **Addition of HCN to aldehydes and ketones**

  $$C_2H_5OH \rightarrow CH_3CHO \rightarrow CH_3CH(OH)CN \rightarrow CH_3CH(OH)COOH$$

  Step 1: warm ethanol with dilute sulphuric acid and potassium dichromate solution and distil off the ethanal as it forms

  Step 2: add HCN and KCN or HCN and a trace of base

  Step 3: hydrolyse by heating under reflux with dilute sulphuric acid.

- **Substitution with KCN**

  $$C_2H_5OH \rightarrow C_2H_5I \rightarrow C_2H_5CN \rightarrow C_2H_5CH_2NH_2$$

  Step 1: add moist red phosphorus and iodine (or convert to chloro or bromo compound)

  Step 2: warm under reflux with KCN in aqueous ethanolic solution

  Step 3: reduce with $NaBH_4$ in aqueous solution.

- **Use of a Grignard reagent** – see page 11 (for Edexcel candidates only).

Analysis – chemical methods

Test	Observation	Conclusion
add bromine water	red-brown colour disappears	>C=C< group present
add bromine water	goes colourless and a white precipitate forms	phenol (not AQA)
warm, under reflux, with dilute NaOH, acidify with dilute HNO_3 and then add silver nitrate solution)	white precipitate soluble in dilute NH_3;	chloro compound
	cream precipitate soluble in conc. but not dilute NH_3;	bromo compound
	pale yellow precipitate not soluble in conc. NH_3	iodo compound
add PCl_5 to dry compound	steamy fumes of HCl	–OH group (acid or alcohol) present
warm with ethanoic acid and a trace of conc. H_2SO_4	sweet (fruity) smell	alcohol (R–OH)
warm with ethanol and a trace of conc. H_2SO_4	sweet (fruity) smell	carboxylic acid (–COOH)
add $NaHCO_3$ or Na_2CO_3 solid or solution	fizzes and gives off CO_2 (lime water test)	carboxylic acid (–COOH)
add 2,4-dinitrophenyl hydrazine (Brady's reagent)	orange precipitate	carbonyl compound, >C=O (aldehyde or ketone)
add ammoniacal silver nitrate (Tollens' reagent)	silver mirror on warming	aldehyde (–CHO)
warm with Fehling's (Benedict's) solution	red precipitate	aldehyde (–CHO)
warm with $K_2Cr_2O_7$(aq) and dilute H_2SO_4	orange solution goes green	primary or secondary alcohol or aldehyde

Analysis – % composition

- See Chapter 2 of *Do Brilliantly – AS Chemistry*, page 19.

Analysis – Spectroscopic methods

- **Infra-red**

Look for:

1 A peak at around 1720 cm^{-1} which indicates a C=O group as in carbonyl compounds, acids and esters.

2 A broad peak (due to hydrogen bonding) between 2500 and 3300 cm^{-1}, caused by the O–H group in acids, and between 3230 and 3550 cm^{-1}, caused by the O–H group in alcohols and phenols.

- **Mass spectra**

Look for:

1 The formation of the molecular ion M^+. Its m/z (m/e) value gives the molar mass.

2 Fragments from the molecular ion: typical fragments are those with an m/e = (M – 15) due to loss of CH_3, CH_3^+ (m/e =15) and other carbocations, and RCO^+. Aromatic compounds sometimes produce $C_6H_5^+$ fragments, but $C_4H_3^+$ is also common.

- **NMR spectra**

Look for:

1 The relative peak heights. These tell you the number of hydrogen atoms in a particular environment, e.g. the number of CH_3 hydrogen atoms, or the number of aromatic CH hydrogen atoms in the molecule. $(CH_3)_3CH$ will have two peaks of relative heights 9:1.

2 The splitting of the peaks. Hydrogen atoms on a neighbouring (adjacent) carbon atom cause a splitting:

(a) A **doublet** is caused by **1 H atom** on a neighbouring carbon atom.

(b) A **triplet** is caused by **2 H atoms** on a neighbouring carbon atom.

(c) A **quartet** is caused by **3 H atoms** on a neighbouring carbon atom.

(d) A **line is not split** if there is **no hydrogen** on the **adjacent carbon** atom or if the **neighbouring** hydrogen atoms are in an **identical group**. Thus, in benzene, the line is not split as all the hydrogens are in identical environments. In $CH_3CH_2CH_2CH_3$, the line due to the CH_2 group is split into a quartet by the CH_3 but is not affected by the neighbouring (identical) CH_2 group. The line due to the CH_3 group is split into a triplet, because of the two hydrogen atoms on the adjacent carbon atom. The ratio of the peak heights of the CH_3 line to the CH_2 line is 3:2 (6 × (CH_3) H atoms to 4 × (CH_2) H atoms).

Note that the **H in OH** always gives a **single** line as it is **hydrogen bonded**.

3 The chemical shift, δ. H atoms in different environments have different chemical shifts.

Questions to try

Examiner's hints *for question 1*
- NaBH$_4$ is a reducing agent. Concentrated sulphuric acid is a dehydrating agent; what type of compound loses water?
- The molar mass of **A** is 86; what fragment must have been lost to give $m/z = 57$?
- In how many different environments are the hydrogen atoms, as only two NMR peaks are observed?
- For question 1 you were supplied with a data table and a NMR spectrum. The spectrum is shown below.
  Look up infra-red absorption frequencies in this chapter.

Q1

Compound **A**, $C_5H_{10}O$, reacts with NaBH$_4$ to give **B**, $C_5H_{12}O$. Treatment of **B** with concentrated sulphuric acid yields compound **C**, C_5H_{10}. Acid-catalysed hydration of **C** gives a mixture of isomers, **B** and **D**.

Fragmentation of the molecular ion of **A**, $[C_5H_{10}O]^{+\cdot}$, leads to a mass spectrum with a major peak at m/z 57. The infra-red spectrum of compound **A** has a strong band at 1715 cm^{-1} and the infra-red spectrum of compound **B** has a broad absorption at 3350 cm^{-1}. The proton n.m.r. spectrum of **A** has two signals at δ 1.06 (triplet) and 2.42 (quartet), respectively.

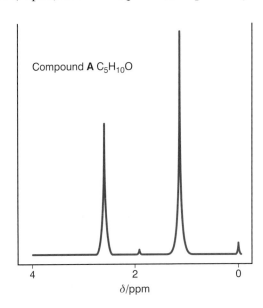

Compound **A** $C_5H_{10}O$

δ/ppm

Use the analytical and chemical information provided to deduce structures for compounds **A**, **B**, **C** and **D**, respectively. Include in your answer an equation for the fragmentation of the molecular ion of **A** and account for the appearance of the proton n.m.r. spectrum of **A**. Explain why the isomers **B** and **D** are formed from compound **C**.

[20 marks]

Questions to try

These are typical synoptic-style questions and you will be required to answer them in the answer booklet, not on the question paper.

Q2

A hydrocarbon **E** has the following composition by mass: C, 90.56%; H, 9.44%. $M_r = 106$

(a) (i)   Use the data above to show that the empirical formula is C_4H_5

  (ii) Deduce the molecular formula.

[3 marks]

(b) Compound **E** contains a benzene ring. Draw structures for all **four** possible isomers of **E**.

[4 marks]

(c) The n.m.r. spectrum of **E** is shown below.

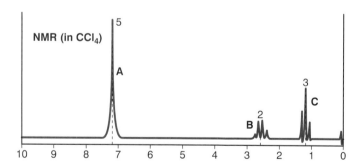

Suggest the identity of the protons responsible for the groups of peaks **A**, **B** and **C**. For each group of peaks, explain your reasoning carefully in terms of both the chemical shift value and the splitting pattern.

[9 marks]

(d) Using the evidence from the peaks in **(c)**, identify and show hydrocarbon **E**.

[1 mark]

[Total 17 marks]

The answers to these questions are on pages 85–86.

3 Chemical equilibrium

Before you start reading this chapter, you should revise Chapter 8 of *Do Brilliantly – AS Chemistry.*

Exam Questions and Student's Answers

1 (a) Write an expression for K_p for each of the following equilibria, giving the units in each case.

(i) $N_2O_4(g) \rightleftharpoons 2NO_2(g)$

$$K_p = \frac{[NO_2]^2}{[N_2O_4]} \text{ ✗ } \text{mol dm}^{-3} \text{ ✗}$$

[2 marks] (0/2)

(ii) $CaCO_3(s) \rightleftharpoons CaO(s) + CO_2(g)$

$$K_p = \frac{(p \text{ of } CaO)(p \text{ of } CO_2)}{(p \text{ of } CaCO_3)} \text{ ✗ } \text{atm ✓}$$

[2 marks] (1/2)

(b) (i) With reference to the equilibrium in **(a)(i)**, calculate the equilibrium partial pressures of N_2O_4 and NO_2 at 60 °C and 1.2 atm pressure, given that 81% of the initial N_2O_4 is dissociated at this temperature.

	N_2O_4	NO_2
Moles at start	1	0
Change	−0.81	+ (2 X 0.81) = 1.62
Moles at eq.	1 − 0.81 = 0.19	1.62 ✓ Total moles = 1.81 ✓
^ Partial pressures/atm	0.19 ✗ X 1.2 ✓	1.62 ✗ X 1.2

[5 marks] (3/5)

(ii) Calculate K_p at this temperature

$$K_p = \frac{1.62}{0.19} \text{ ✗ } = 8.5 \text{ atm}$$

[1 mark] (0/1)

(c) The equilibrium reaction

$2SO_2(g) + O_2(g) \rightleftharpoons 2SO_3(g)$ $\Delta H = -196 \text{ kJ mol}^{-1}$

is used in the manufacture of sulphuric acid and uses vanadium (V) oxide as the catalyst.

(i) Write the expression for K_c for this equilibrium

$$K_c = \frac{[SO_3]^2}{[SO_2]^2[O_2]} \text{ ✓}$$

[1 mark] (1/1)

(ii) What effect does the catalyst have on the position of equilibrium in this reaction?

A catalyst has no effect on the position of equilibrium. ✓ It speeds up both the forward and the back reactions equally.

[1 mark] ①/₁

(iii) A steel vessel of volume 2.0 dm³ has introduced into it 0.20 mol of SO_3, 0.040 mol of SO_2 and 0.010 mol of O_2. By calculation of the **apparent** value of K_c, show that this mixture is **not at equilibrium**, and explain in which direction the system will move in order to achieve equilibrium at a temperature of 800 K.

The value of K_c at this temperature is 1.7×10^6 mol⁻¹ dm³.

$$K_c \text{ (apparent)} = \frac{0.20^2}{0.040^2 \times 0.010} \, \cancel{X} = 2500, \text{ which is}$$

not equal to K_c and so the system is not in equilibrium ✓ ∧

[3 marks] ①/₃

[Total 15 marks]

⑦/₁₅

2 Hydrogen and iodine react in a reversible reaction according to the equation

$$H_2(g) + I_2(g) \rightleftharpoons 2HI(g) \qquad \qquad \Delta H = -10.4 \text{ kJ mol}^{-1}$$

(a) A vessel of volume 4.0 dm³ was filled with 0.10 mol of hydrogen and 0.10 mol of iodine and the mixture was heated to a temperature of 730 K. The value of K_c at this temperature is 49. Calculate the concentration of hydrogen iodide at equilibrium.

	H_2	I_2	HI
Moles at start	0.10	0.10	0

Let x mol of HI be formed

Change	$-x/2$	$-x/2$	$+x$
Moles at equilibrium	$(0.10 - x/2)$	$(0.10 - x/2)$	x ✓
Concentration at eq.	$(0.10 - x/2) \div 4$	$(0.10 - x/2) \div 4$	$x \div 4$ ✓

$$K_c = \frac{[HI]^2}{[H_2].[I_2]} \checkmark = \frac{(x/4)^2}{(0.10 - x/2)/4 \cdot (0.10 - x/2)/4} = 49$$

The 4's cancel. Square rooting both sides ✓

$$\frac{x}{(0.10 - x/2)} = 7 \text{ or } x = 0.7 - 3.5x \text{ or } 4.5x = 0.7$$

or $x = 0.7/4.5 = 0.1556$

The concentration of HI = 0.1556 mol/4.0 dm³ = 0.0388888 ✗ mol dm⁻³

[5 marks] ④/₅

23

(b) Explain the effect, if any, on the value of K_c and hence on the position of equilibrium of:

(i) increasing the temperature

The position shifts in the endothermic direction, and so K_c decreases. ✔ ✗

[3 marks] $\frac{1}{3}$

(ii) increasing the pressure

K_c will not ✔ change in value as it is only altered by a change in temperature, and the position of equilibrium will not alter either, because there are the ✔ same number of gas molecules on both sides of the equation.

[2 marks] $\frac{2}{2}$

$\frac{7}{10}$ [Total 10 marks]

How to score full marks

1 (a) (i) You must not confuse K_c, which has equilibrium concentrations in its expression, **with K_p,** which is in terms of partial pressures. The correct answer is:

$$K_p = \frac{(\text{p of } NO_2)^2}{(\text{p of } N_2O_4)}$$

and the **units** are atm (or kPa or **any pressure unit**).

(ii) The **partial pressure of solids is ignored** in the expression for K_p. The answer should be:

$$K_p = (\text{p of } CO_2)$$

and the **units** are atm (or kPa or any **pressure unit**).

(b) (i) The student set out the calculation very clearly, but then made the error of not **working out the mole fraction** of each substance. The **correct answer** is:

	N_2O_4	NO_2	
Moles at start	1	0	
Change	−0.81	+ (2 × 0.81) = 1.62	
Moles at eq.	1 − 0.81 = 0.19	1.62	Total moles = 1.81
Mole fraction	0.19/1.81 = 0.105	1.62/1.81 = 0.895	
Partial pressures/atm	0.105 × 1.2 = 0.126	0.895 × 1.2 = 1.07	

(ii) The student forgot that the **partial pressure** of NO_2 **must be squared.** The correct response is:

$$K_p = \frac{(1.07)^2}{0.126} = 9.1 \text{ atm}$$

(c) (iii) The student made a common error by using **moles** instead of **concentration** in the expression for apparent K_c. The mark for realising that the system is not in equilibrium was gained, but the student made another common error of **not reading the question fully**. It asked for an **explanation** of the direction in which the system would move, and this was omitted. The **correct answer** is:

The apparent K_c (the quotient) $= \dfrac{(0.20/4)^2}{(0.040/4)^2 . (0.010/4)} = 10000 \text{ mol}^{-1}\text{dm}^3 < 1.7 \times 10^6$

This is not equal to K_c, and so the system is not at equilibrium.

As the apparent K_c (the quotient) is less than K_c, its value must get bigger. So the top line must increase, which means that more SO_3 must be made, and the system will move to the right.

2 (a) This was very well answered by the student, who set out the calculation clearly but made one mistake. The student gave the final **answer to 6 significant figures, when only 2 are justified**. The final answer should have been **0.039 mol dm^{-3}**.

(b) (i) There are two errors here. The student had the explanation the wrong way round. It is the **change in the value of K_c which causes the position of equilibrium to move**. Also the student failed to state which is the **endothermic direction**. The correct answer is: **As the reaction is exothermic left to right, K_c will decrease. This means that the position of equilibrium will move to the left.**

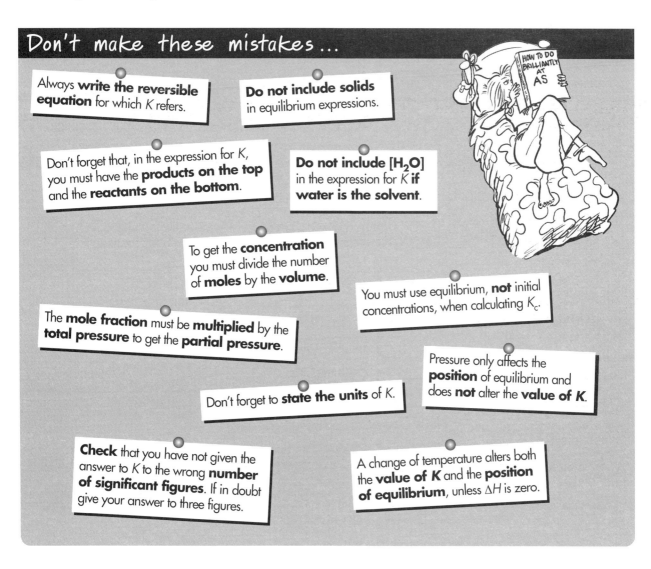

Don't make these mistakes...

Always **write the reversible equation** for which K refers.

Do not include solids in equilibrium expressions.

Don't forget that, in the expression for K, you must have the **products on the top** and the **reactants on the bottom**.

Do not include [H$_2$O] in the expression for K **if** water is the solvent.

To get the **concentration** you must divide the number of **moles** by the **volume**.

You must use equilibrium, **not** initial concentrations, when calculating K_c.

The **mole fraction** must be **multiplied** by the **total pressure** to get the **partial pressure**.

Pressure only affects the **position** of equilibrium and does **not** alter the **value of K**.

Don't forget to **state the units** of K.

Check that you have not given the answer to K to the wrong **number of significant figures**. If in doubt give your answer to three figures.

A change of temperature alters both the **value of K** and the **position of equilibrium**, unless ΔH is zero.

Key points to remember

The **concentration** of a substance is the **number of moles divided by the volume** (usually in dm^3).

For a **reaction** $nA + mB \rightleftharpoons xC + yD$, where n, m, x and y are the numbers of A, B, C and D in the equation,

the quotient $\dfrac{[C]^x.[D]^y}{[A]^n.[B]^m} = K_c$ at equilibrium.

If the quotient does not equal K_c, the system is not at equilibrium.

K_c should be calculated in a table:

moles at start	Usually given as data
moles at start	Usually given as data
change of moles	Use the stoichiometry of the equation
moles at equilibrium	One value is normally given in the data

$[\ ]_{eq} = \dfrac{\text{moles at equilibrium}}{\text{volume}}$

- The **partial pressure** of a gas in a mixture of gases is defined as its mole fraction times the total pressure.
- The **mole fraction** is the number of moles of that substance divided by the total number of moles.

- For **heterogeneous reactions involving gases**, the concentration or partial pressures of any solid is ignored. For example, for the reaction:

$Fe_2O_3(s) + 3CO(g) \rightleftharpoons 2Fe(s) + 3CO_2(g)$

$K_c = \dfrac{[CO_2]^3}{[CO]^3}$ and $K_p = \dfrac{(\text{partial pressure of } CO_2)^3}{(\text{partial pressure of } CO)^3}$

- K_c and K_p have **units**, which must **always** be stated.

For a **gaseous reaction** $nA(g) + mB(g) \rightleftharpoons xC(g) + yD(g)$, where n, m, x and y are the numbers of A, B, C and D in the equation, the quotient

$\dfrac{(\text{partial pressure of } C)^x (\text{partial pressure of } D)^y}{(\text{partial pressure of } A)^n (\text{partial pressure of } B)^m}$

$= K_p$ at equilibrium.

If the quotient does not equal K_p, then the system is not at equilibrium.

Calculation of K_p is more difficult than K_c, but should also be done **in a table**:

moles at start	Usually given as data
moles at start	Usually given as data
change of moles	Use the stoichiometry of the equation
moles at equilibrium	One value is normally given in the data
total moles at equilibrium	This is the sum of all the moles at equilibrium

mole fraction $= \dfrac{\text{moles of each}}{\text{total moles}}$

partial pressure = mole fraction $\times$ total pressure

Effect of factors on the values of K_c and K_p and on the position of equilibrium:

- Temperature: an **increase of temperature** will **decrease** the value of K for all **exothermic** reactions, and will **increase** the value of K for **endothermic** reactions. It will also drive the **position of equilibrium** in the **endothermic** direction.

- Pressure: an **increase in pressure** has **no effect** on the value of K, but it will drive the **position of equilibrium** towards the side with **fewer gas molecules**.

- Catalyst: a catalyst has **no effect** on either the value of K or the position of equilibrium, but will **reduce the time taken** to reach equilibrium.

Questions to try

Examiner's hints *for question 1*

(a) You will need to work out whether the reaction is exothermic or endothermic. Remember that the question is about the effect on K_p, not on the position of equilibrium.

(b) Don't forget to calculate the mole fraction of each and hence the partial pressures.

Q1

Ethane can be cracked at high temperature to yield ethene and hydrogen according to the equation:

$$C_2H_6(g) \rightleftharpoons C_2H_4(g) + H_2(g)$$

The standard enthalpy of formation of ethene is positive, whereas that of ethane is negative.

(a) Discuss the effect on the equilibrium constant, K_p, of changes to

(i) the temperature

....

....

(ii) the pressure.

....

....

[3 marks]

(b) Calculate the value of the equilibrium constant, K_p, for this cracking reaction, given that 1.00 mol of ethane under an equilibrium pressure of 180 kPa at 1000 K can be cracked to produce an equilibrium yield of 0.36 mol of ethene.

[7 marks]

[Total 10 marks]

Q2

Ethanol can be manufactured either from starch or from ethene. In the latter process, ethene is hydrated by passing it and steam over a phosphoric acid catalyst at a temperature of 275 °C and a pressure of 70 atm (7000 kPa). The reaction is:

$$C_2H_4(g) + H_2O(g) \rightleftharpoons C_2H_5OH(g) \qquad\qquad \Delta H = -46\,kJ\,mol^{-1}$$

(a) Explain why the conditions stated for the manufacturing process are the most economic.

....

....

....

....

....

[5 marks]

(b) Suggest how the unreacted ethene can be separated from the equilibrium mixture, and what is then done with it.

....

....

[2 marks]

(c) When 1.00 mol of ethene and 1.00 mol of steam are mixed in a vessel of volume 3.00 dm^3 and allowed to reach equilibrium, 90.0% of the ethene reacts. State the expression for K_c and calculate its value.

[5 marks]

The answers to these questions are on page 87.   [Total 12 marks]

4 Acid–base equilibria

Exam Question and Student's Answer

1 (a) (i) Define pH

pH is the log of the hydrogen concentration. ✗

[1 mark] 0/1

(ii) Define the term 'weak acid' as applied to methanoic acid, HCOOH.

A weak acid is partially ionised. ✓
∧

[2 marks] 1/2

(b) Calculate the pH of the following solutions. (The ionic product of water, $K_w = 1.00 \times 10^{-14}$ mol^2 dm^{-6} at 25 °C; the acid dissociation constant for methanoic acid is 1.78×10^{-4} mol dm^{-3}):

(i) a solution of hydrochloric acid of concentration 0.152 mol dm^{-3}

$[H^+] = 0.152$ ✓ mol dm^{-3} $pH = -\log (0.152) = 0.82$ ✓

[2 marks] 2/2

(ii) a solution of sodium hydroxide of concentration 0.747 mol dm^{-3}

$[OH^-] = 0.747$ ✓ mol dm^{-3} $pH = -\log (0.747) = 0.13.$ ✗

[3 marks] 1/3

(iii) a solution of methanoic acid of concentration 0.152 mol dm^{-3}

∧ $[H^+] = [HCOO^-]$ ✓

$K_a = \sqrt{K_a.[HCOOH]} = \sqrt{(1.78 \times 10^{-4} \times 0.152)}$ ✓

$= 5.20 \times 10^{-3}$ mol dm^{-3}

$pH = 2.2839$ ✗

[4 marks] 2/4

(c) (i) What is the principal property of a buffer solution?

It has a constant pH ✗ *when* ∧ *acid or alkali is added*

[2 marks] 0/2

(ii) The acid dissociation constant, K_a, for ethanoic acid is 1.80×10^{-5} mol dm^{-3}. Calculate the pH of a buffer solution which has a concentration of 0.105 mol dm^{-3} with respect to ethanoic acid and 0.342 mol dm^{-3} with respect to sodium ethanoate.

$$K_a = 1.80 \times 10^{-5} = \frac{[H^+].[CH_3COO^-]}{[CH_3COOH]} = \frac{[H^+].[salt]}{[weak\ acid]} \checkmark$$

$$[H^+] = 1.8 \times 10^{-5} \times \frac{0.105}{0.342} = 5.526 \times 10^{-6}\ mol\ dm^{-3}$$

$$pH = -log\ (5.526 \times 10^{-6}) = 5.26 \checkmark$$

[3 marks] 3/3

[Total 17 marks] 10/17

2 This question is about propanoic acid, $C_2H_5CO_2H$.

(a) K_a for propanoic acid is 1.3×10^{-5} mol dm^{-3}.

(i) Complete the equation below for the dissociation of propanoic acid.

$C_2H_5CO_2H(aq) + H_2O(l) \rightleftharpoons$ $C_2H_5CO_2^-(aq) + H_3O^+(aq) \checkmark$

[1 mark] 1/1

(ii) Write the expression for K_a:

$$K_a = \frac{[H_3O^+].[C_2H_5CO_2^-]}{[C_2H_5CO_2H]\ [H_2O]}\ \times$$

[1 mark] 0/1

(iii) Calculate the pH of a 0.10 M solution of the acid. Only an approximate calculation is required.

$$[H_3O^+] = [C_2H_5CO_2^-] \qquad [C_2H_5CO_2H] = 0.10\ (approx.) \checkmark$$

$$[H_3O^+]^2 = K_a\ [C_2H_5CO_2H] = 1.3 \times 10^{-5} \times 0.10$$

$$[H_3O^+] = 1.14 \times 10^{-3} \quad pH = 2.94 \checkmark$$

[2 marks] 2/2

(b) (i) In the titration of a solution of propanoic acid with sodium hydroxide solution, what would you expect the approximate value of the pH to be equal to when equal numbers of moles of the acid and alkali had reacted? No further calculation is required.

Approximate value of pH: it will have a pH of 7 ✗

Justify your answer: because all the acid has been neutralised ✗

[2 marks] 0/2

30

(ii) Name a suitable indicator you could use to determine the end point in this titration.

✓
phenolphthalein

[1 mark] $\frac{1}{1}$

(c) (i) The pH of a buffer solution is given by the equation

$$pH = -\log K_a - \log [acid]_{eqm}/[base]_{eqm}$$

What ratio of propanoic acid solution and sodium propanoate, both 0.100 m, would be needed to make a buffer solution of pH 4.2? Give your answer to an appropriate number of significant figures.

$4.2 = -\log(1.3 \times 10^{-5}) - \log [acid]/[base]$

$\log[acid]/[base] = 4.89 - 4.2 = 0.69$ ✓

$[acid]/[base] = 10^{-0.69} = 0.20$ ✗

[2 marks] $\frac{1}{2}$

(ii) Use your answer to **(i)** to calculate the volume of the propanoic acid solution you would need to mix with the sodium propanoate solution to make 1 dm³ of this buffer.

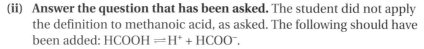
As you need 0.2 times as much acid, you must add 0.2 dm³ of acid. ✗

[1 mark] $\frac{0}{1}$

[Total 10 marks] $\frac{5}{10}$

How to score full marks

1 **(a) (i)** The answer is wrong. pH is **minus** log of the hydrogen ion concentration.

(ii) **Answer the question that has been asked.** The student did not apply the definition to methanoic acid, as asked. The following should have been added: $HCOOH \rightleftharpoons H^+ + HCOO^-$.

Sometimes this equilibrium is written $HCOOH + H_2O \rightleftharpoons H_3O^+ + HCOO^-$. **You can write either H^+ or H_3O^+ for the 'hydrogen ion'.**

(b) (ii) The student made a common error, and should have realised that **NaOH is a base**, so its pH must be >7. The first step is correct, but $-\log[OH^-] = pOH$. The correct calculation is: $[OH^-] = 0.747$, then

either: $pOH = -\log(0.747) = 0.13$, then $pH = 14 - 0.13 = 13.87$

or: $[H^+] = 1.00 \times 10^{-14}/[OH^-] = 1.339 \times 10^{-14}$ mol dm⁻³

$pH = -\log(1.339 \times 10^{-14}) = 13.87$

(iii) Start with the expression for K_a, which is: $K_a = [H^+][HCOO^-]/[HCOOH]$. The second error is that the answer has too many numbers. **Always give pH values to two decimal places.**

(c) (i) There are two common errors here. **Buffers do not have a constant pH** and only work when **small** amounts of acid or alkali are added. The correct answer is: 'a buffer solution resists change in pH when small amounts of acid or alkali are added'.

2 (a) (ii) If you have H_2O **in the equation**, you must **not** include $[H_2O]$ in your expression for K_a because water is the solvent, and so its **concentration does not alter.**

(b) (i) The student made a common error. The pH at the end (equivalence) point of a titration depends upon the **strengths of the acid and the base**. It is only 7 if **both** are strong. The correct answer is: 'the pH will be about 9, as sodium propanoate is the salt of a weak acid (propanoic) and a strong base (sodium hydroxide)'.

(c) (i) The student confused this calculation with the one where you have to calculate $[H^+]$ from pH when, as pH = **minus** log $[H^+]$, $[H^+] = 10^{-pH}$. In this calculation 0.69 equals **plus** log [acid]/[base], so the ratio is $10^{+6.9}$.

(ii) The question asked for a total volume of 1 dm^3. The calculation is:

Let volume of acid = $x\,dm^3$, then volume of NaOH = $(1 - x)\,dm^3$

$x/(1 - x) = 4.9$; $5.9x = 4.9$; volume of acid = $x = 0.83\,dm^3$ ($830\,cm^3$).

Don't make these mistakes...

The **more acidic** the solution, the **lower** the pH. Solutions of **weak acids** have a pH of **around 3**.

Strong acids are **totally** ionised ($HCl \rightarrow H^+ + Cl^-$), but **weak** acids are **partially** ionised ($HA \rightleftharpoons H^+ + A^-$)

The **end** (or equivalence) **point** of a titration is **not** at a pH of 7, unless it is a **strong** acid/ strong base titration.

Do **not** put $[H_2O]$ into expressions for K_a as water is the solvent.

The value of K_w **changes with temperature**, so at **high** temperatures neutral pH is **less than 7**.

Lowry–Brønsted acid/base conjugate pairs are connected by the loss or gain of an H^+ ion.

Use the **log not the ln** button on your calculator. To go from pH to $[H^+]$ you have to use the 2nd log buttons.

A buffer solution does not have a **constant** pH; it **resists** a change in pH when **small** amounts of H^+ or OH^- ions are added.

Phenolphthalein is used for titrations involving **weak acids**, methyl orange for titrations involving **weak bases** and **either** for **strong acid/strong base** titrations.

Acids, bases and pH

- **Acids** produce H^+ ions in solution; **alkalis** produce OH^- ions.

- A **Lowry–Brønsted acid gives a proton** (H^+ ion) to a base, and a **Lowry–Brønsted base accepts a proton** (H^+ ion) from an acid. The species formed from an acid by the **loss of a proton** is called the acid's **conjugate base**. For example, in the reaction:

  $$H_2SO_4 + HNO_3 \rightleftharpoons H_2NO_3^+ + HSO_4^-,$$

  HSO_4^- is the conjugate base of the acid H_2SO_4 and $H_2NO_3^+$ is the conjugate acid of the HNO_3 which here is acting as a base.

- A **strong acid** is **totally** ionised in solution.

- A **weak acid** is only **partially** ionised in solution. For the weak acid HA:

  $$HA \rightleftharpoons H^+ + A^- \qquad K_a = [H^+].[A^-]/[HA]$$

pH, pOH and pK

- **pH of a solution = –log [H⁺]** and **pK_a of a weak acid = –log K_a**.
  Therefore $[H^+] = 10^{-pH}$ and $K_a = 10^{-pK}$

- In all aqueous solutions $[H^+] \times [OH^-] = K_w = 1.00 \times 10^{-14}$ at 25 °C.
  Therefore pH + pOH = 14 where pOH = –log [OH⁻].

- In all neutral solutions $[H^+] = [OH^-]$. At 25 °C, $[H^+] = 1.00 \times 10^{-7}$ mol dm⁻³.

- At 25 °C **neutral** solutions have a **pH of 7**. In **acidic** solutions **[H⁺] > [OH⁻]** and so **pH <7**. In **alkaline** solutions **[OH⁻] > [H⁺]** and so **pH >7**.

- The pH of a **strong acid** solution = **–log (concentration of the acid)**. For example, the pH of a 0.123 mol dm⁻³ solution of HCl = –log (0.123) = 0.91. **Always express pH values to 2 decimal places.**

- The pOH of a 0.123 mol dm⁻³ solution of a **Group 1 hydroxide** (e.g. NaOH) is –log (0.123) = 0.91. Therefore the pH = 14 – 0.91 = 13.09

- The pOH of a 0.123 mol dm⁻³ solution of a **Group 2 hydroxide** (e.g. Ba(OH)₂) = –log (**2** × 0.123) = 0.61. Therefore the pH = 14 – 0.61 = 13.39

- To calculate the **pH** of a solution of a **weak** acid, you must know the **concentration** and K_a for the acid. For example, the pH of a 0.123 mol dm⁻³ solution of a weak acid HA, with $K_a = 2.2 \times 10^{-5}$ mol dm⁻³ is given by:

  $$HA \rightleftharpoons H^+ + A^- \qquad K_a = [H^+].[A^-]/[HA]$$

  $[H^+] = [A^-]$, and $[HA] = 0.123$ mol dm⁻³

  $2.2 \times 10^{-5} = [H^+]^2/0.123$: $\qquad [H^+]^2 = 0.123 \times 2.2 \times 10^{-5} = 2.706 \times 10^{-6}$

  $[H^+] = \sqrt{(2.706 \times 10^{-6})} = 0.00164$; $\qquad$ pH = –log (0.00164) = 2.78

Buffer solutions

- A **buffer** solution **resists** changes in pH when **small** amounts of acid or base are added.
- All buffers are a **mixture** of a weak acid and its salt – e.g. ethanoic acid and sodium ethanoate (or a weak base and its salt).
- To calculate the **pH of a buffer** solution, you must know the **concentration** of the weak acid and of its salt as well as K_a for the acid. For example: the pH of a buffer solution made by mixing 0.10 mol of a weak acid HA, of $K_a = 3.4 \times 10^{-6}$, with 0.20 moles of its salt, NaA, all in $1.0\,dm^3$ of solution, is given by:

  $$NaA \rightarrow Na^+ + A^- \quad \text{and } HA \rightleftharpoons H^+ + A^- \quad K_a = [H^+] \times [A^-]/[HA]$$

  $$[A^-] = [salt] = 0.20/1.0\,mol\,dm^{-3}$$

  $$[HA] = [weak\ acid] = 0.10/1.0\ mol\ dm^{-3}$$

  $$[H^+] = K_a\,[HA]/[A^-] = K_a\,[weak\ acid]/[salt] = 3.4 \times 10^{-6} \times 0.10/0.20 = 1.7 \times 10^{-6}$$

  $$pH = -\log(1.7 \times 10^{-6}) = 5.77$$

- A buffer solution resists a change in pH because when H^+ ions are added they are **removed** by the reaction $H^+ + A^- \rightarrow HA$. The A^- ions come from the large reservoir of A^- **ions** from the **salt** NaA, which is **totally ionised**: $NaA \rightarrow Na^+ + A^-$.
- When OH^- ions are added to a buffer, they are **removed** by reaction with the large reservoir of the **HA** molecules from the weak **acid**: $OH^- + HA \rightarrow H_2O + A^-$.

Titrations

- The **end point** of a titration is when the **relative number of moles** according to the equation have been added – e.g. for the reaction between HCl and NaOH (a 1:1 reaction), the end point occurs when an equal number of moles of each have been added.
- There are **three types** of titration:

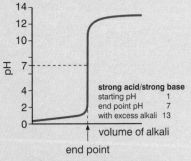

1 When a **strong base** (NaOH) is added to a **strong acid** (HCl)

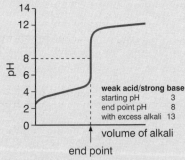

2 When a **strong base** is added to a **weak acid**.

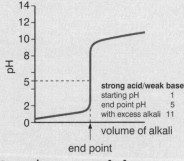

3 When a **weak base** is added to a **strong acid**.

Choice of indicator

- Choose an indicator whose pK_{ind} value is well within the vertical part of the graph.
- For a **strong acid/strong base** titration, choose **methyl orange**, $pK_{ind} = 3.7$ or **phenolphthalein**, $pK_{ind} = 9.3$
- For a **weak acid/strong base**, choose **phenolphthalein**.
- For a **strong acid/weak base**, choose **methyl orange**.

Questions to try

Examiner's hints *for question 1*

(a) Look carefully at the formula for barium hydroxide before calculating [OH$^-$].

(b) Ask yourself what the starting pH, the equivalence point pH and the pH after an excess of barium hydroxide has been added will be. Also, what volume of barium hydroxide will be needed to reach the equivalence point?

(c) Make sure that you label both pairs.

(d) What is the relationship between [H$^+$] and [OH$^-$] in a neutral solution?

Q1

(a) Calculate the pH of a 0.123 mol dm^{-3} solution of the strong base barium hydroxide, $Ba(OH)_2$.

....

[2 marks]

(b) **(i)** Sketch the curve showing how the pH of 20 cm^3 of a 0.123 mol dm^{-3} solution of the weak acid ethanoic acid varies as 25 cm^3 of a 0.123 mol dm^{-3} solution of barium hydroxide is steadily added.

[4 marks]

(ii) Select which of the following indicators would be the most suitable for this titration. Explain your choice.

Indicator	methyl red	bromothymol blue	cresol red
pK_{ind}	5	7	8

....

[2 marks]

(c) In the esterification reaction of ethanoic acid with ethanol in the presence of concentrated sulphuric acid catalyst, the first step is:

$$CH_3CO_2H + H_2SO_4 \rightarrow CH_3CO_2H_2^+ + HSO_4^-$$

Identify the Lowry–Brønsted acid/base conjugate pairs, indicating clearly which is the acid and which the base in each pair.

....

[2 marks]

(d) K_w for the ionisation of water is defined as $K_w = [H^+][OH^-]$. Its value at a temperature of $40\,°C$ is $3.8 \times 10^{-14}\,mol^2\,dm^{-6}$. Calculate the pH of a neutral solution at this temperature.

[2 marks]

[Total 12 marks]

Examiner's hints for question 2
(a) The opposite of strong is weak **not** dilute.
(b) Use the ratio of NaOH to H_3PO_4 from the equation to convert moles H_3PO_4 to moles NaOH.

Q2

This question refers to different aspects of acid/base chemistry:

(a) Hydrochloric acid, HCl, is classified as a **strong** acid but it can have both **concentrated** and **dilute** solutions. Explain why this is so.

....

....

[3 marks]

(b) Sodium phosphate, Na_3PO_4, a water softening agent, can be prepared in the laboratory by neutralising phosphoric acid.

A student prepared this compound in the laboratory from $20.0\,cm^3$ of $0.100\,mol\,dm^{-3}$ phosphoric acid and $0.250\,mol\,dm^{-3}$ sodium hydroxide:

$$H_3PO_4(aq) + 3NaOH(aq) \rightarrow Na_3PO_4(aq) + 3H_2O(l)$$

(i) Deduce the oxidation state of phosphorus in sodium phosphate, Na_3PO_4.

....

[1 mark]

(ii) Calculate the volume of NaOH(aq) that the student would need to use to just neutralise the phosphoric acid, using the quantities above.

[3 marks]

(c) Calculate the pH of the NaOH(aq) used in (b) ($K_w = 1.00 \times 10^{-14}\,mol^2\,dm^{-6}$).

[4 marks]

The answers to these questions are given on page 88.

[Total 11 marks]

5 | Kinetics

Before you start reading this chapter, you should revise Chapter 7 of *Do Brilliantly – AS Chemistry*.

Exam Question and Student's Answer

1 Hydrogen reacts with iodine at 450°C to give hydrogen iodide. The results from several experiments designed to find the rate equation for the reaction are given below.

Initial $[I_2]$/mol dm^{-3}	Initial $[H_2]$/mol dm^{-3}	Relative initial rate
0.001	0.001	1
0.003	0.001	3
0.001	0.004	4

(a) (i) Find the order of reaction with respect to each of the reactants.

The order with respect to iodine is 1, ✓ and for hydrogen is also 1. ✓

[3 marks] (2/3)

(ii) Write the rate equation for the reaction.

The rate = $[H_2].[I_2]$ ✗

[1 mark] (0/1)

(iii) Explain why the rate equations cannot be written from the stoichiometric (chemical) equation for the reaction, but must be found experimentally.

The chemical equation tells you the ratio of moles that react and does not tell you anything about the mechanism. ✓ The rate equation contains only those species whose concentrations affect the rate of reaction. ✓

[2 marks] (2/2)

(iv) What do you understand by the **rate-determining step** in a chemical reaction?

It is the step that controls the rate of reaction. ✗

[1 mark] (0/1)

(v) Illustrate your answer to **(iv)** by writing out any reaction mechanism of your choice.

[3 marks] (2/3)

(b) Reaction rates generally increase with temperature for two reasons.

    (i) State what these reasons are.

The activation energy increases with temperature. ✗

The collision frequency increases as the temperature rises. ✓

[2 marks] ①/₂

    (ii) Sketch on the axes a graph of the distribution of molecular energies at a given temperature T_1 and at a higher temperature T_2, and hence explain the increase in rate.

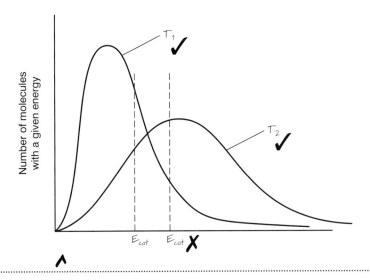

[4 marks] ②/₄

    (iii) By use of the graph, or otherwise, explain the effect of a catalyst on the rate of a chemical reaction.

The activation energy for the route with a catalyst is less than

that without a catalyst. ✓

[2 marks] ①/₂

10/18   [Total 18 marks]

How to score full marks

1 (a) (i) Explanations must always be given when working out orders of reaction. The student should have written: 'When the [I_2] is multiplied by 3 and the [H_2] not altered, the rate also increases by a factor of 3, so the order with respect to I_2 is 1. Likewise, when [H_2] is multiplied by 4 and [I_2] is unaltered, the rate increases by 4 times. So the order with respect to H_2 is also 1'.

(ii) The student forgot to include the **rate constant**, k. The equation is: **Rate = k [I_2]. [H_2]**

(iv) Don't simply restate the question, which is what the student did. **It will never score any marks.** The answer is: 'It is the **slowest** step in a reaction, and hence controls the overall rate'.

(v) Any 2 (or more) step reaction will answer this question. Here the student chose an S_N1 reaction and gave a good account of the mechanism, but failed to state which is the **slower** step. The phrase 'the first step is the slower and hence is the rate determining step' should have been added.

(b) (i) The student made a common error. **Temperature does not alter the value of the activation energy.** The correct answer is: 'Because the average kinetic energy of the molecules increases, the number with energy $\geq E_a$ increases, so more collisions result in reaction.' The second reason was correctly stated.

(ii) The student made the error of putting the E_a line to the left of the T_2 peak. **Activation energies are always to the right,** unless the reaction is extremely rapid. The student also forgot to **explain why the rate is increased**. The correct answer is: 'At the higher temperature more molecules, on collision, have energy greater than or equal to the activation energy'.

(iii) The student omitted to **explain** why a lower activation energy results in a faster reaction. The answer should have continued: 'therefore more molecules will have energy $\geq E_{cat}$'.

Don't make these mistakes...

Do **not** use the **equation** to work out the **order of reaction**; you must use the **experimental data**.

Don't forget that an **increase in temperature** causes an **increase in the rate constant**, and that **fast** reactions have **smaller activation energies** and **larger rate constants** than slower reactions.

Don't forget to **give your reasons** when calculating the order of reaction from initial rate data.

The rate-determining step is the **slowest** step in the series of steps. The order of a substance involved only **after** the rate-determining step will be zero.

Remember that the **rate** of all chemical reactions **increases** as the **temperature rises** and that this increase is **because a greater number of colliding molecules** have **energy** greater than the activation energy.

A rate constant has **units**, which depend on the total order of reaction.

A catalyst doesn't lower the activation energy; it provides an **alternative route** which has a lower activation energy. A catalyst does **not** affect the **enthalpy change** or the **position of equilibrium**.

Key points to remember

- The **overall** order of a reaction is the **sum** of the orders with respect to all the reactants.

- The order with respect to a particular reagent can be worked out from initial rates as long as only the concentration of one reagent is altered.

- The order of a reaction **must be found experimentally** and can then be used to suggest a mechanism.

- The activation energy is the **minimum** energy that the colliding molecules must have in order for the collision to result in reaction.

- The **larger** the value of the **activation energy**, the **smaller** the value of the **rate constant** and the **slower the reaction**.

- The rate-determining step is the **slowest** step in the mechanism, and hence has the largest value of activation energy.

- An **increase in temperature** will **increase** the value of the **rate constant**.

- A **catalyst** provides an alternative path with a **lower E_a** and hence a **larger** value of the **rate constant**.

- The **half-life** of a first-order reaction is constant at a given temperature. The half-life is defined as the **time for the concentration of the reactant to halve**.

Maxwell-Boltzmann distribution

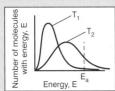

This shows the way in which the kinetic energy of the molecules in a gas (or liquid) are distributed at a temperature T_1. The number of molecules that have energies greater than or equal to a value E_a is equal to the area under the curve to the right of the E_a line.

At a higher temperature T_2, a larger number have energies greater than or equal to this value.

Factors that affect the rate of reaction

Factor	How it affects the rate	Why it affects the reaction
Temperature	An increase in temperature increases the rate of a chemical reaction. This is true for both exothermic and endothermic reactions.	The particles have more kinetic energy so more of them have energy $\geq$ the activation energy. Thus more of them react on collision. The collision frequency also increases but slightly (see diagram above).
Concentration	An increase in the concentration of substances in solution increases the rate of reaction.	The particles are closer together and so the frequency of collision is increased.
Pressure	An increase in pressure of a gas increases the rate of reaction.	As above, the gas molecules are closer together and so the frequency of collision is increased.
Catalyst	A catalyst speeds up a reaction without getting used up.	A catalyst provides an alternative reaction route which has a lower activation energy, thus, at a given temperature, more particles will have kinetic energy greater than or equal to the activation energy.
Surface area	The rate of a reaction of a solid with a gas or a liquid is dependent on the surface area of the solid – the bigger the surface area the faster the reaction.	The frequency of collision of a reactant molecule or ion with the surface is increased.

Questions to try

Examiner's hints *for question 1*

(a) **(i)** You must show how you obtained the order of reaction for each gas.

(ii) Select the data from any one of the experiments. Don't forget the units of the rate constant, k.

(b) You must explain why your chosen mechanism gives the correct order for both reactants.

Q1

(a) Nitrogen dioxide and carbon monoxide react according to the equation

$$NO_2(g) + CO(g) \rightarrow NO(g) + CO_2(g).$$

The kinetics of this reaction were studied and the following results were obtained:

Experiment	$[NO_2]$/mol dm^{-3}	[CO]/mol dm^{-3}	Rate of reaction/mol dm^{-3} s^{-1}
1	1.2×10^{-2}	3.4×10^{-3}	1.3×10^{-5}
2	2.4×10^{-2}	3.4×10^{-3}	5.2×10^{-5}
3	1.2×10^{-2}	6.8×10^{-3}	1.3×10^{-5}

(i) Deduce the rate equation for the reaction.

....

....

....

[5 marks]

(ii) Calculate the value of the rate constant.

[2 marks]

(b) Two possible mechanisms have been suggested for this reaction. They are:

I   a single-step reaction: $NO_2 + CO \rightarrow NO + CO_2$

II   a two-stage reaction: *step 1*. $NO_2 + NO_2 \rightarrow NO + NO_3$ (slow)

                                        step 2. $NO_3 + CO \rightarrow NO_2 + CO_2$ (fast)

Which mechanism is supported by the rate equation that you obtained in **(a)**?

....

....

[2 marks]

(c) For the two steps in the two-stage mechanism, state and explain:

(i)   which has the larger value of activation energy

....

[1 mark]

(ii)   which has the larger value of rate constant.

....

[1 mark]

[Total 11 marks]

Examiner's hints for question 2
- Remember your answer to **(a)(iii)** when answering **(b)**.
- Do not confuse the reaction in **(a)**, involving chemicals X and Y, with the reaction in **(b)**, which involves chemicals A and B.

Q2

(a) A chemical reaction is first order with respect to compound **X** and second order with respect to compound **Y**.

(i)   Write the rate equation for this reaction.

....

[2 marks]

(ii)   What is the overall order of this reaction?

....

[1 mark]

(iii) By what factor will the rate increase if the concentrations of **X** and **Y** are **both** doubled?

....

....

[1 mark]

(b) The table below shows the initial concentrations of two compounds, **A** and **B**, and also the initial rate of the reaction that takes place between them at constant temperature.

Experiment	[A]/mol dm^{-3}	[B]/mol dm^{-3}	Initial rate/mol dm^{-3} s^{-1}
1	0.2	0.2	3.5×10^{-4}
2	0.4	0.4	1.4×10^{-3}
3	0.8	0.4	5.6×10^{-3}

(i) Determine the overall order of the reaction between **A** and **B**. Explain how you reached your conclusion.

Overall order of reaction: ....

Explanation: ....

[2 marks]

(ii) Determine the order of reaction with respect to the compound B. Explain how you reached your conclusion.

Order with respect to B: ....

Explanation: ....

[2 marks]

(iii) Write the rate equation for the overall reaction.

....

[1 mark]

(iv) Calculate the value of the rate constant, stating its units.

....

....

[2 marks]

The answers to these questions are given on page 89.   [Total 11 marks]

Before you read this chapter, you will find it helpful to revise Chapter 6 in *Do Brilliantly – AS Chemistry*.

Exam Questions and Student's Answers

1 (a) Using the following data, construct a Born–Haber cycle for potassium chloride and use it to calculate the electron affinity of chlorine.

	ΔH/kJ mol^{-1}
1st ionisation energy of potassium	+419
Enthalpy of atomisation of potassium	+ 89.2
Enthalpy of atomisation of chlorine	+121.7
Enthalpy of formation of KCl(s)	–436.7
Lattice enthalpy of potassium chloride	–711

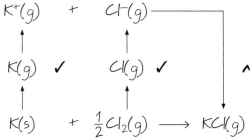

The electron affinity of chlorine = -355.6 kJ mol^{-1} ∧

[5 marks] ③/5

(b) Calcium is in the same period in the periodic table as potassium. The lattice enthalpy of calcium chloride is –2258 kJ mol^{-1}. Explain why this is so different from the value for potassium chloride given in **(a)**.

There is a greater electrostatic force of attraction between the positive and negative ions in CaCl$_2$ than in KCl, because the calcium ion is 2+, whereas the potassium ion is only 1+. ✓ ∧

[2 marks] ½

(c) Lattice enthalpies may be calculated based on an assumption about the structure of the solid or found experimentally using data in the Born–Haber cycle. The experimental lattice enthalpy of potassium chloride is 9 kJ mol^{-1} more exothermic than that calculated; for calcium chloride the value is 35 kJ mol^{-1} more exothermic than calculated.

Suggest why the calculated and experimental values are different in both compounds.

The Born–Haber ✗ value assumes that the solid is 100% ionic. KCl is almost 100% ionic so the difference is very small, ✓ but CaCl$_2$ is significantly covalent, ∧ and so the difference is much larger.

[3 marks] ⅓

(d) The solubility of calcium sulphate in water at room temperature is much greater than that of barium sulphate. Suggest reasons for this difference.

The Ba^{2+} ion is about the same size as the SO_4^{2-} ion ✗ and so

barium sulphate's lattice enthalpy is larger ✗ than calcium sulphate's.

[2 marks] 0/2

[Total 12 marks]

5/12

2 (For Nuffield and AQA candidates **only**)

(a) Predict the sign of ΔS for the reaction and ΔS for the surroundings in the reaction:

$$N_2(g) + 3H_2(g) \rightarrow 2NH_3(g) \qquad\qquad \Delta H = -92.4 \text{ kJ mol}^{-1}$$

The reaction is exothermic, therefore it is feasible, ✗ and so ΔS for

the reaction is positive. ✗

[2 marks] 0/2

(b) Calculate the temperature at which the reaction $2FeO(s) + C(s) \rightarrow 2Fe(s) + CO_2(g)$ becomes feasible (spontaneous), given the following data:

$\Delta H_{formation}$/kJ mol^{-1}: FeO(s) = –272; CO_2(g) = –394.

S/J K^{-1} mol^{-1}:   FeO(s) = +61; CO_2(g) = +214;

Fe(s) = +27;   C(s) = +6

$\Delta H_{reaction} = \Delta H_f$ products $- \Delta H_f$ reactants $= (-394) - (2 \times -272) = +150$ kJ ✓

$\Delta S_{reaction} = \Delta S$ products $- \Delta S$ reactants $= (2 \times 27 + 214) - (2 \times 61 + 6)$

$= +141$ JK^{-1} ✓

The reaction becomes feasible when $\Delta G = 0$, which is when $\Delta H = T\Delta S$ ✓

$T = \Delta H/\Delta S = 150/141$ ✗ $= 1.06$ °C ✗

[5 marks] 3/5

(c) The reduction of silver oxide by carbon is thermodynamically feasible, at room temperature Explain why there is no reaction at room temperature.

As the reaction is thermodynamically feasible, ΔG must be negative ✓

(or $-\Delta H/T + \Delta S_{system}$ must be > 0), but the reaction is not

observed because it is too slow. ^

[2 marks] 1/2

5/9   [Total 9 marks]

1 **(a)** The student drew the correct Born–Haber cycle but failed to put values or symbols for each step. Another mark was lost for not giving the working in the calculation. You **must show your working** to get full marks; it also allows the examiner to give you some marks if you make a careless mathematical error. A good answer would be:

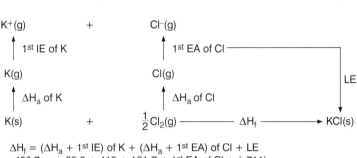

$$\Delta H_f = (\Delta H_a + 1^{st} \text{ IE}) \text{ of K} + (\Delta H_a + 1^{st} \text{ EA}) \text{ of Cl} + \text{LE}$$
$$-436.7 = +89.2 + 419 + 121.7 + 1^{st} \text{ EA of Cl} + (-711)$$
$$1^{st} \text{ EA of Cl} = -436.7 - 89.2 - 419 - 121.7 + 711 = -355.6 \text{ kJ mol}^{-1}$$

(b) The student made the first point well, but failed to **give a second reason** why calcium chloride's lattice enthalpy is larger. It is: 'a calcium ion is also smaller than a potassium ion, which also make the electrostatic force stronger.'

(c) It is the **calculated** value which is based on the assumption that the substance is 100% ionic. Also the student gave an incomplete answer, as it should have included an **explanation** as to **why calcium chloride is less ionic** (more covalent). The reason is that 'the calcium ion, being more highly charged and smaller, polarises the chloride ion, making the bond partially covalent.'

(d) **The answer given is quite wrong.** A Ba^{2+} ion is much smaller than an SO_4^{2-} ion, and lattice enthalpies **always decrease down a group**. The correct answer is: 'The value of –(the lattice enthalpy) + the hydration enthalpies of the ions determines the solubility. The lattice enthalpy and the hydration enthalpy of the cation both decrease down a group, but the lattice enthalpy decreases more because the SO_4^{2-} ion is so much bigger than either cation.'

2 **(a)** The answer is totally wrong. **You have to break it down into two parts.** Firstly 'because the reaction is exothermic, $\Delta S_{surroundings}$ is positive.' Secondly, you have to make a statement about $\Delta S_{reaction}$ (ΔS_{system}). This is 'because the reaction goes from 4 gas molecules to 2 gas molecules, $\Delta S_{reaction}$ is negative (the system becomes **less** disordered)'.

(b) The student's answer was excellent until the last line, when the common error was made of not realising that ΔH **is measured in kJ** and ΔS **in J**. The second error was that T in the expression $T = \Delta H/\Delta S$ must be in **kelvin** not °C. The correct last line is: '$T = \Delta H/\Delta S = 150 \times 10^3 \text{J}/141 \text{J K}^{-1} = 1060 \text{ K (or } 787°\text{C)}$.'

(c) The student failed to say **why** the reaction is too slow. The answer is 'because the **activation energy** for the reaction is too high for enough molecules to have this energy at room temperature'.

Don't make these mistakes...

Don't go into the exam on this unit not knowing how to construct a **Born–Haber cycle**.

Don't confuse the enthalpy of atomisation with enthalpy of dissociation. The former is for the reaction producing 1 mol of atoms and the other for 2 mol of atoms.

Don't get the signs wrong. For instance, –(lattice enthalpy) is a positive number, as all lattice enthalpies are negative.

Key points to remember

Born–Haber cycle

- The **enthalpy of atomisation**, ΔH_a, of an element is the enthalpy change to **make 1 mol** of **gaseous** atoms, e.g. Na(g) or Cl(g), from the element in its stable state – e.g. from Na(s) or $Cl_2(g)$. For chlorine it is for $\frac{1}{2}Cl_2(g) \rightarrow Cl(g)$, which is half the (bond) dissociation energy.
- The **1st ionisation** energy, 1st IE, is the energy required to **remove one electron** from a mole of gaseous atoms, e.g. for sodium it is for: $Na(g) \rightarrow Na^+(g) + e^-$.
- The **energy change** for $Mg(s) \rightarrow Mg^{2+}(g) + 2e^-$ is the **sum** of the **1st** and the **2nd** ionisation energies.
- The **electron affinity**, EA, is the energy change when **one electron is added** to a mole of gaseous atoms, e.g. for chlorine it is for $Cl(g) + e^- \rightarrow Cl^-(g)$
- The **lattice enthalpy**, LE, is the enthalpy change when the **separate gaseous ions form 1 mol of crystalline solid**, e.g. for sodium chloride it is for: $Na^+(g) + Cl^-(g) \rightarrow NaCl(s)$.
- Note that Edexcel, Nuffield and OCR always define lattice enthalpies (energies) as the **formation** of the lattice, whereas AQA accept either the enthalpy change for the formation or the dissociation of the lattice. AQA candidates must make it clear which definition they are using.
- The Born–Haber cycle for MgO is:

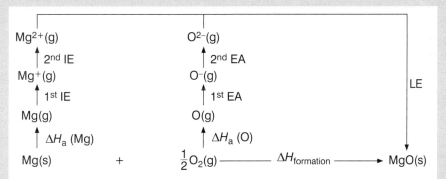

- ΔH_f of MgO(s) = (ΔH_a + 1st IE + 2nd IE) of Mg + (ΔH_a + 1st EA + 2nd EA) of O + LE

Factors affecting lattice enthalpy

- The **bigger the charge** on either ion, the **larger** (more exothermic) the value of the **lattice enthalpy**.
- The **smaller the ions**, the **larger** the value of the **lattice enthalpy**.
- The **less the % ionic character**, the **larger** the value of the **lattice enthalpy**.
- Certain ionic radii ratios, e.g. ions of dissimilar size, cause a slight lowering of the lattice enthalpy.

Solubility of ionic solids

- $\Delta H_{solution}$ is the enthalpy change for 1 mol of solid dissolving in excess water.
- $\Delta H_{solution} = -LE + \Delta H_{hydration}$ of the cation $+ \Delta H_{hydration}$ of the anion.
- $\Delta H_{hydration}$ is the enthalpy change when 1 mol of a gaseous ion dissolves in water.
- The **more exothermic**, or the **less endothermic**, the value of $\Delta H_{solution}$, the **more soluble** is the solid.
- Both the **lattice enthalpy** and the **hydration enthalpy** of the cation **decrease** down a group in the periodic table.
- The **solubility** of the group 2 hydroxides **increases** down the group because the lattice enthalpy decreases more than the hydration energy decreases. This is because the **cation** and the **anion** are of **similar size**.
- The group 2 sulphates **decrease** in **solubility** down the group because the lattice enthalpy decreases less than the hydration energy decreases. This is because the **sulphate ion is much bigger** than any of the group 2 cations.

Entropy   (AQA and Nuffield candidates only)

- Entropy, symbol S, is a measure of the **disorder** of a system.
- ΔS for a **solid producing a liquid or gas** is **positive** (more disordered).
- ΔS for a **liquid producing a gas** is also positive.
- ΔS for a **solid dissolving** is positive
- An **exothermic** reaction (ΔH negative) causes the entropy of the **surroundings** to increase.
- $\Delta S_{surroundings} = q/T$, where q is the heat produced by the reaction
- **(AQA)** The change in free energy, ΔG, is given by: $\Delta G = \Delta H - T\Delta S$.
  If ΔG is **negative**, the reaction will be **spontaneous**. If ΔG is **positive**, the reaction **will not happen** (not spontaneous). A reaction changes from being spontaneous to not being spontaneous when $\Delta G = 0$, that is when $\Delta H = T\Delta S$.
- **(Nuffield)** A reaction changes from being spontaneous to not being spontaneous when $\Delta S_{surroundings} + \Delta S_{system} = 0$. As $\Delta S_{surroundings} = q/T = -\Delta H/T$. Change-over occurs when $\Delta H = T\Delta S$.
- **(AQA and Nuffield)** The feasibility of a reaction can be predicted from ΔS and ΔH values:

ΔH	ΔS	Feasibility
negative	positive	always
negative	negative	if $\Delta H > T\Delta S$
positive	negative	never
positive	positive	if $T\Delta S > \Delta H$

Examiner's hints *for question 1*
(a) Draw the cycle using all the data. Remember that ΔH_f is for the reaction forming 1 mol of RbCl from its elements.
(b) Li and Rb ions have the same charge. In what way do they differ?

The lattice enthalpy of rubidium chloride, RbCl, can be determined indirectly using a Born–Haber cycle.

(a) Use the data in the table below to construct the cycle and to determine a value for the lattice enthalpy of rubidium chloride.

[6 marks]

enthalpy change	energy/kJ mol^{-1}
formation of rubidium chloride	−435
atomisation of rubidium	+81
atomisation of chlorine	+122
1st ionisation energy of rubidium	+403
1st electron affinity of chlorine	−349

(b) Explain why the lattice enthalpy of lithium chloride, LiCl, is more exothermic than that of rubidium chloride.

....

....

[2 marks]

[Total 8 marks]

Examiner's hints *for question 2*
(a), (b) and (d) Start by writing the chemical equation for the process. Then calculate ΔS. Remember the conditions for change-over of spontaneity, and so calculate ΔH, for (a), or T, for (d).
(c) Whether a reaction is observed at a certain temperature depends on both thermodynamics and kinetics.

Q2

(For Nuffield and AQA candidates only)

Use the data in the table below to answer the following questions. Give chemical equations and calculate numerical values of ΔS wherever possible.

(a) At all temperatures below 100°C, steam at atmospheric pressure condenses spontaneously to form water. Explain this observation in terms of ΔG and calculate the molar enthalpy for the process $H_2O(g) \longrightarrow H_2O(l)$ at 100°C.

....

....

[6 marks]

(b) Explain why the reaction of 1 mol of methane with steam to form carbon monoxide and hydrogen ($\Delta H^{\ominus} = +210$ kJ mol^{-1}) is spontaneous only at high temperatures.

....

....

[6 marks]

(c) Explain why the change of 1 mol of diamond to graphite ($\Delta H^{\ominus} = -2$ kJ mol^{-1}) is feasible at all temperatures yet does not occur at room temperature.

....

[3 marks]

(d) The reaction between 1 mol of calcium oxide and carbon dioxide to form calcium carbonate ($\Delta H^{\ominus} = -178$ kJ mol^{-1}) ceases to be feasible above a certain temperature, T_s. Determine the value of T_s.

....

[2 marks]

Entropy data

Species	$S^{\ominus}$/J K^{-1} mol^{-1}	Species	$S^{\ominus}$/J K^{-1} mol^{-1}
C (graphite)	6	$H_2O(g)$	189
C (diamond)	3	$H_2O(l)$	70
$H_2(g)$	131	$CH_4(g)$	186
$CO(g)$	198	$CaO(s)$	40
$CO_2(g)$	214	$CaCO_3(s)$	90

The answers to these questions are on page 90.   [Total 17 marks]

7 The periodic table – period 3

Before you read this chapter, you will find it helpful to revise Chapter 4 in *Do Brilliantly – AS Chemistry*.

Exam Questions and Student's Answers

1 (a) Complete the following table:

	Na	**Mg**	**Al**	**Si**
Formula of anhydrous chloride	$NaCl$	$MgCl_2$ ✓	Al_2Cl_6	$SiCl_4$ ✓

[2 marks]

(b) (i) Write equations, including state symbols, for the changes that take place when these chlorides are added to water.

sodium chloride $\quad NaCl \rightarrow Na^+ + Cl^-$ ∧

magnesium chloride $\quad MgCl_2(s) + H_2O \rightarrow MgO + 2HCl$ ✗

aluminium chloride $\quad Al_2Cl_6 + 2H_2O(l) \rightarrow_∧ Al(OH)_3(s) + 6HCl (aq)$

silicon chloride $\quad SiCl_4(l) + 2H_2O \rightarrow SiO_2(s) + 4HCl(aq)$ ✓

[4 marks]

(ii) Interpret these changes on the basis of the bonding in the chlorides.

Sodium chloride and magnesium chloride are ionic, whereas aluminium chloride and silicon chloride are covalent. ✓ ∧

[2 marks]

(iii) Suggest why the chloride of carbon, CCl_4, does not react with water.

Carbon ∧ cannot accept the lone pair of electrons from water ✓ and the four chlorine atoms are so big that water molecules are sterically hindered from attacking the central carbon atom. ✓

[4 marks]

(c) Aluminium hydroxide is **amphoteric**. Write two **ionic** equations to show the meaning of this statement.

$Al(OH)_3(s) + 3H^+(aq) + 3Cl^-(aq) \rightarrow Al^{3+}(aq) + 3Cl^-(aq) + 3H_2O(l)$ ✗

$Al(OH)_3(s) + 3OH^-(aq) \rightarrow [Al(OH)_6]^{3-}(aq)$ ✓

[2 marks]

[Total 14 marks]

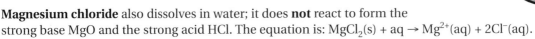

(b) (i) **Sodium chloride:** the student forgot to **put state symbols** in the equation.
The correct equation is: $NaCl(s) + aq \rightarrow Na^+(aq) + Cl^-(aq)$.

Magnesium chloride also dissolves in water; it does **not** react to form the strong base MgO and the strong acid HCl. The equation is: $MgCl_2(s) + aq \rightarrow Mg^{2+}(aq) + 2Cl^-(aq)$.

Aluminium chloride: the student forgot to **balance the equation.** The right-hand side should be: $\rightarrow 2Al(OH)_3(s) + 6HCl(aq)$

(ii) The student stated the bonding, but **did not draw the conclusion** from the bond type. A good answer is: 'sodium chloride and magnesium chloride are ionic and so just dissolve in water, but aluminium chloride and silicon chloride are covalent and so are hydrolysed to acidic solutions.'

(iii) The student failed to **explain why** carbon cannot accept a pair of electrons, and why this would cause the C—Cl bonds to break. A good answer would be: 'Carbon does not have any d orbitals which could accept a pair of electrons. Thus the C—Cl bond would have to break before the new C—O bond forms and this is energetically unlikely. Also the four chlorine atoms sterically hinder the approach of a water molecule and prevent a reaction.'

(c) **An ionic equation must not contain any spectator ions** – these are ions which are exactly the same on both sides of the equation. So the Cl^- ions must not be in the equation, which should be:
$Al(OH)_3(s) + 3H^+(aq) \rightarrow Al^{3+}(aq) + 3H_2O(l)$

Don't make these mistakes ...

Remember, sodium burns in **oxygen** to form a **peroxide**, Na_2O_2, not an oxide.

Sodium reacts with **water** to form hydrogen and sodium **hydroxide**, but **magnesium** reacts with **steam** to form hydrogen and magnesium **oxide**.

Don't get the formulae of the chlorides and oxides of the period 3 elements **wrong**.

Anhydrous aluminium chloride is **covalent**, but **hydrated** aluminium chloride, $AlCl_3.6H_2O$, is ionic.

The formula of **solid** anhydrous aluminium chloride is **AlCl₃**, but when vaporised it forms Al_2Cl_6 molecules.

Covalent chlorides of period 3 elements react with water to form HCl, and so the solution is acidic. **Don't forget that acidic solutions have a low pH.**

Ionic chlorides just **dissolve** in water – they **do not react** with it.

REVISION NOTES

Key points to remember

Oxides

- The period 3 elements react with **oxygen** to form: Na_2O_2 (sodium peroxide), MgO, Al_2O_3, SiO_2, P_2O_3 or with excess oxygen P_2O_5 and SO_2.
- Sodium oxide, Na_2O, also exists. Like the other metal oxides in this period it is **ionic**.
- The oxides of the **non-metals** are **covalent**.
- Na_2O and MgO are both **bases** and **react with acids** to form **ionised salts**:
  e.g. $MgO(s) + 2H^+(aq) \rightarrow Mg^{2+}(aq) + H_2O(l)$
- Al_2O_3 is **amphoteric**, which means that it reacts as a base **and** as an acid:
  $Al_2O_3(s) + 6H^+(aq) \rightarrow 2Al^{3+}(aq) + 3H_2O(l)$

  $Al_2O_3(s) + 6OH^-(aq) + 3H_2O(l) \rightarrow 2[Al(OH)_6]^{3-}(aq)$

- The **non-metal oxides** are all **acidic**, but SiO_2 is so weak that it only reacts with molten NaOH: $SiO_2(s) + 2NaOH(l) \rightarrow Na_2SiO_3(l) + H_2O(g)$
- P_2O_5 reacts with **water** to form phosphoric(V) acid, which is partially ionised:
  $P_2O_5 + 3H_2O \rightarrow 2H_3PO_4 \rightleftharpoons 2H^+ + 2H_2PO_4^-$
- SO_2 reacts with water to form sulphuric (IV) acid (**sulphurous** acid), which is also partially ionised: $SO_2 + H_2O \rightarrow H_2SO_3 \rightleftharpoons H^+ + HSO_3^-$

Chlorides

- The **formulae** of the period 3 chlorides are: $NaCl(s)$, $MgCl_2(s)$, $AlCl_3(s)$ or $Al_2Cl_6(g)$ or $AlCl_3.6H_2O(s)$, $SiCl_4(l)$, $PCl_3(l)$ or $PCl_5(s)$. (The chlorides of sulphur are complex.)
- **Sodium** and **magnesium** chlorides are **ionic**. **Anhydrous** aluminium chloride is covalent, but the **hydrated** chloride is **ionic**. All the **non-metal** chlorides are **covalent**.
- **Sodium** and **magnesium** chlorides **dissolve** in water: e.g. $NaCl(s) + aq \rightarrow Na^+(aq) + Cl^-(aq)$
- **Hydrated** aluminium chloride is **acidic** in solution because the high charge density on the Al^{3+} ion causes an H^+ ion to be lost: $[Al(H_2O)_6]^{3+}(aq) \rightleftharpoons H^+(aq) + [Al(H_2O)_5OH]^{2+}(aq)$
- Anhydrous aluminium chloride and silicon chloride react with water to form an oxide and hydrochloric acid: e.g. $SiCl_4(l) + 2H_2O(l) \rightarrow SiO_2(s) + 4H^+(aq) + 4Cl^-(aq)$
- **Phosphorus** chlorides react with water to form an acid of phosphorus and hydrochloric acid:
  e.g. $PCl_5(s) + 4H_2O(l) \rightarrow H_3PO_4(aq) + 5H^+(aq) + 5Cl^-(aq)$

Elements + water

- **Sodium** reacts **violently** with **cold** water to form a **hydroxide** and hydrogen:
  $2Na(s) + 2H_2O(l) \rightarrow 2Na^+(aq) + 2OH^-(aq) + H_2(g)$
- **Magnesium** and **aluminium** react when **heated** with **steam** to form an **oxide** and hydrogen: e.g. $2Al(s) + 3H_2O(g) \rightarrow Al_2O_3(s) + 3H_2(g)$
- **Silicon, phosphorus** and **sulphur** do **not** react with water.
- **Chlorine** reacts **reversibly** with water to form a strong and a weak acid:
  $Cl_2(g) + H_2O(l) \rightleftharpoons H^+(aq) + Cl^-(aq) + HOCl(aq)$

Questions to try

Q1

(a)   Write equations to show what happens when the following oxides are added to water and predict approximate values for the pH of the resulting solutions

   (i) sodium oxide

Equation..   pH ....

[2 marks]

   (ii) sulphur dioxide

Equation..   pH ....

[2 marks]

(b)   What is the relationship between bond type in the oxides of the Period 3 elements and the pH of the solutions which result from addition of the oxides to water?

....

....

[2 marks]

(c)   Write equations to show what happens when the following chlorides are added to water and predict approximate values for the pH of the resulting solutions

   (i) magnesium chloride

Equation..   pH ....

[2 marks]

   (ii) silicon tetrachloride

Equation..   pH ....

[2 marks]

The answers to this question are on page 91.   [Total 10 marks]

8 Redox equilibria

Before reading this chapter, it would be helpful to revise Chapter 5 of *Do Brilliantly – AS Chemistry*.

Exam Questions and Student's Answers

1 The data given below are taken from the electrochemical series.

Reaction at 298K	$E^\ominus$/V
$MnO_4^{2-}(aq) + 4H^+(aq) + 2e^- \rightarrow MnO_2(s) + 2H_2O$	+1.55
$MnO_4^-(aq) + 8H^+(aq) + 5e^- \rightarrow Mn^{2+}(aq) + 4H_2O$	+1.51
$MnO_4^-(aq) + e^- \rightarrow MnO_4^{2-}(aq)$	+0.60

Disproportionation is the term used for a reaction in which an element changes from a single oxidation state to two different oxidation states, one being higher and the other lower than the original. This can be illustrated by the reaction

$$Cl_2(aq) + H_2O(l) \rightarrow 2H^+(aq) + Cl^-(aq) + OCl^-(aq)$$

in which Cl(0) becomes Cl(–1) and Cl(+1).

Use the information given above, wherever relevant, to answer the questions that follow.

(a) What is meant by the term *electrochemical series*?

It is a list of metals according to their reactivity. ✗

[1 mark] 0/1

(b) On warming, OCl^- ions change into Cl^- and ClO_3^-. Write an equation for this disproportionation reaction and determine the oxidation state of chlorine in the ClO_3^- ion.

Equation: $2OCl^-(aq) \rightarrow Cl^-(aq) + ClO_3^-(aq)$ ✗

Oxidation state in ClO_3^- ion: $Cl + (3 \times -2) = -1$

Oxidation state of Cl is +5 ✓

[2 marks] 1/2

(c) How does the overall redox potential show that a reaction is spontaneous?

If the overall $E^\ominus$ value is greater than 0.3 volts ✗

[1 mark] 0/1

(d) The manganate(VI) ion, MnO_4^{2-}, is also unstable and undergoes spontaneous disproportionation to form manganate(VII) ions and solid manganese(IV) oxide. Construct an overall equation for this disproportionation and use values from the electrochemical series to calculate the overall $E^\ominus$ value for the reaction.

$MnO_4^{2-}(aq) + 4H^+(aq) + 2e^- \rightarrow MnO_2(s) + 2H_2O(l)$ ✓ $E^\ominus = +1.55$ V

$2MnO_4^{2-}(aq) \rightarrow 2MnO_4^-(aq) + 2e^-$ ✓ $\qquad\qquad E^\ominus = 2 \times 0.60$ V ✗

so overall equation is

$3MnO_4^{2-}(aq) + 4H+(aq) + 2e^- \rightarrow MnO_2(s) + 2MnO_4^-(aq) + 2H_2O(l) + 2e^-$ ✗

Overall $E^\ominus = +1.55 + 1.2 = 2.75$ V

[4 marks] 2/4

3/8 [Total 8 marks]

2 Iron forms an ion of formula FeO_4^-, which, in acid solution, will oxidise Sn^{2+} ions to Sn^{4+} ions.

(a) Calculate the oxidation number of iron in the FeO_4^- ion. ✓

$Fe + 4 \times (-2) = -1$: Oxidation number $= +7$

[1 mark] 1/1

(b) A titration found that $25.0\,cm^3$ of a $0.107\,mol\,dm^{-3}$ solution of tin(II) chloride required $21.3\,cm^3$ of a solution containing $0.0502\,mol\,dm^{-3}$ of FeO_4^- ions for complete oxidation. Calculate the ratio of moles of Sn^{2+} ions to FeO_4^- ions in the redox reaction.

moles of $Sn^{2+} = 0.107 \times 25 = 2.675$ ✗

moles of $FeO_4^- = 0.0502 \times 21.3 = 1.069$ (✓)

Ratio moles $Sn^{2+}: FeO_4^- = 2.675/1.069 = 2.50$ or $5:2$ ✓

[3 marks] 2/3

(c) Hence calculate the change in oxidation number of the iron and the formula of the iron ion formed when FeO_4^- ions are reduced.

Change in oxidation number of Sn^{2+} = change in oxidation number of FeO_4^- ✗

Tin goes up by 2, therefore iron goes down by 2 to +5 ✗

3/7 [3 marks] 0/3

[Total 7 marks]

1 **(a)** The student's definition is wrong on two counts. Firstly the series is **not limited to elements**, and secondly it refers to a **reduction** process. The correct response is: 'It is a series listing redox equations according to their reduction potentials, with the most negative at the top.' (This means that the substance on the left of the top equation in the series is the worst oxidising agent and the substance on the right of the top equation is the best reducing agent.)

(b) Equations must balance as well as show the correct species. **Always check your equations for balancing.** This should be: $3ClO^-(aq) \rightarrow 2Cl^-(aq) + ClO_3^-(aq)$

(c) The correct response is: 'if the overall redox potential is positive (>0), then the reaction will be thermodynamically feasible (spontaneous).' Note that feasibility tells you nothing about the speed of the reaction.

(d) The student correctly wrote the two half equations but, when they were added to get the overall chemical equation, the student forgot to **cancel the electrons**. The correct overall equation is:

$$3MnO_4^{2-}(aq) + 4H^+(aq) \rightarrow MnO_2(s) + 2MnO_4^-(aq) + 2H_2O(l)$$

Also two common errors were made in assigning the value of $E^\ominus$. If a **half equation is reversed**, the sign of **E must be reversed**, but if the equation is **multiplied by an integer** such as 2, the value of E is not **multiplied** by 2. The second half equation with its $E^\ominus$ value should be:

$$2MnO_4^{2-}(aq) \rightarrow 2MnO_4^-(aq) + 2e^- \qquad E^\ominus = - (+0.60\,V) = -0.60\,V$$

and the overall $E^\ominus$ for the reaction $= + 1.55 + (-0.60) = + 0.95\,V$

2 **(b)** When you calculate the number of moles, you must convert the volume to dm^3 by dividing by 1000. The correct answers are:

moles of $Sn^{2+} = 0.107 \times 25.0/1000 = 0.002675$

moles of $FeO_4^- = 0.0502 \times 23.1/1000 = 0.001069$

(c) As the **total** change in oxidation number of the reducing agent will be the same as the **total** change in oxidation number of the oxidising agent, you must first work out by how much the **five** Sn^{2+} ions change. This equals the change of the iron atoms in the **two** FeO_4^- ions. The correct answer is: 'The total change in oxidation number (number of atoms x change of each) of tin equals the total change in the oxidation number of the iron atoms. As the ratio is $5Sn^{2+}$ to $2\,FeO_4^-$, and each Sn^{2+} changes by +2, the total change for the two Fe atoms is –10, or –5 each. Thus the final oxidation state of iron is $+7 - 5 = +2$.'

Don't make these mistakes ...

Don't forget state symbols in redox equations.

Don't quote the 'anticlockwise rule' without writing out the two half equations in the **correct order**.

Don't forget to **reverse the sign** of E if you reverse the half equation.

Remember that **half equations always contain electrons** and that overall equations do not.

In redox (and acid/base) titrations, you must **give the equation**, so that you can **get the stoichiometric ratio correct** when converting moles of one substance to moles of the other.

If you **double a half equation**, **don't** alter the value of E.

Half equations are given as **reduction processes** with electrons on the left, so remember that oxidation agents are on the left in these equations.

In titration calculations remember that all **volumes** must be in dm^3, **not** cm^3.

Make sure all your equations balance. This includes the total changes on the left and right.

Redox titrations

- Estimation of the concentration of a reducing agent (RA)
  Titrate against standard potassium manganate(VII) solution:

  Pipette 25 cm³ of reducing agent, add excess dilute sulphuric acid and titrate against the potassium manganate(VII) solution until a faint pink colour is obtained.

  Write the overall redox equation, then:

$$\frac{\text{moles } KMnO_4}{\text{moles RA}} \left\{ \text{from equation} \right\} = \frac{\text{molarity* } \times \text{ volume (in dm}^3\text{) of } KMnO_4}{\text{molarity } \times \text{ volume (in dm}^3\text{) of RA}}$$

  *the molarity of a solution is the same as its concentration in mol dm⁻³.

- Estimation of the concentration of an oxidising agent (OA)
  Add excess acidified potassium iodide and titrate the liberated iodine against sodium thiosulphate, $Na_2S_2O_3$, solution. Write equations for the liberation of iodine and its reaction with the sodium thiosulphate, and so work out the ratio of moles OA to moles $S_2O_3^{2-}$.

$$\frac{\text{moles OA}}{\text{moles } S_2O_3^{2-}} \left\{ \text{from equations} \right\} = \frac{\text{molarity OA } \times \text{ volume (in dm}^3\text{)}}{\text{molarity } S_2O_3^{2-} \times \text{ volume (in dm}^3\text{)}}$$

- Make sure that you can write half and overall equations, and that you can calculate oxidation numbers (see *Do Brilliantly – AS Chemistry*, Chapter 5).

Cell diagrams
For AQA and Nuffield only

- The convention for writing a cell diagram is:
  Anode | anode compartment || salt **b**ridge || cathode compartment | **c**athode.   (a, b, c)

  Remember that oxidation occurs at the anode (o and a are both vowels) and that reduction occurs at the cathode (both consonants). For the Fe^{3+}/Fe^{2+} and Cl_2/Cl^- system, the Fe^{2+} ions are oxidised, and so the cell diagram is:

  $Pt | Fe^{2+}(aq), Fe^{3+}(aq) || Cl_2(g), Cl^-(aq) | Pt$

- Note that the **reactants** (here Fe^{2+} and Cl_2) are written **first** in each electrode compartment.
- If, after writing a cell diagram, you find that E_{cell} is **negative**, then you have written the diagram **backwards**.

Electrode potentials

- The **standard electrode potential**, $E^{\ominus}$ is defined as *the voltage produced by a reduction reaction at an electrode, measured relative to a standard hydrogen electrode, when all solutions are at a concentration of 1 mol dm^{-3} and all gases are at 1 atm. pressure.*

  For the element zinc, it is the voltage of a zinc electrode dipping in a 1 mol dm^{-3} solution of zinc ions, connected, via a salt bridge, to a standard hydrogen electrode.

- The standard **hydrogen electrode** consists of hydrogen gas at 1 atm. pressure bubbling over a platinum electrode, which is dipping into a 1 mol dm^{-3} solution of H^+ ions. Its half equation is $2H^+(aq) + 2e^- \rightleftharpoons H_2(g)$. Its $E^{\ominus}$ value is defined as zero.

- The overall **electrode potential** (E_{cell} or $E_{reaction}$) can be calculated in various ways:

  From half equations. If you are asked to calculate E_{cell} for the reaction between zinc and copper sulphate solution, you refer to the two half equations, which are:

  $$Zn^{2+}(aq) + 2e^- \rightleftharpoons Zn(s) \qquad E^{\ominus} = -0.76\,V \quad : \quad Cu^{2+}(aq) + 2e^- \rightleftharpoons Cu(s) \qquad E^{\ominus} = +0.34\,V$$

  As zinc and copper ions are the reactants, they must be on the left of the overall equation, so you must reverse the first equation and leave the second as it is. The $E^{\ominus}_{cell} = -(-0.76) + (+0.34) = +1.10\,V$. As the fist equation was reversed, the sign of its $E^{\ominus}$ value has to be reversed too.

  If you have written a cell diagram (see below), you can calculate E_{cell} from:

  $$E^{\ominus} = E_{\text{right-hand electrode}} - E_{\text{left-hand electrode}}.$$

- **Feasibility of a reaction** is predicted by calculating the E_{cell}. If it is >0 (positive), then the reaction is thermodynamically feasible (spontaneous).

$E^{\ominus}_{cell}$ and K

- If $E^{\ominus}_{cell}$ is **positive**, then ΔS_{total} will also be positive.
- The value of K for the reaction can be calculated from $E^{\ominus}_{cell}$ by the expression:

  $$E^{\ominus}_{cell} = \frac{RT \ln K}{n\Im} \qquad$$ where n is the number of electrons exchanged in the redox reaction and $\Im$ is a constant (the Faraday).

  At 25°C this can be written as:

  $$E^{\ominus}_{cell} = \frac{0.0257}{n} \ln K$$

  Thus the bigger the value of $E^{\ominus}_{cell}$, the bigger the value of K.

Questions to try

Q1

(a) (i) Construct the half equations for Fe^{3+} being reduced to Fe^{2+} and for MnO_4^- being reduced in **acid** solution to Mn^{2+}

....

....

[2 marks]

(ii) Hence write the overall ionic equation for the reaction, in acid solution, between Fe^{2+} and MnO_4^- ions.

....

....

[2 marks]

(b) Cast iron is an alloy of carbon and iron. A 4.45 g sample of cast iron was reacted with excess dilute sulphuric acid, and the resulting solution, which contained Fe^{2+} ions, was made up to $250\,cm^3$ with distilled water. $25.0\,cm^3$ portions of this required $23.4\,cm^3$ of a 0.0655 mol dm^{-3} solution of potassium manganate(VII).

Calculate the % purity of the cast iron.

The answers to this question are on page 91.   [5 marks]

9 Transition metals

Exam Question and Student's Answer

1 Copper is a transition metal. A typical property of a transition metal such as copper is the ability to from complex ions with ligands such as water molecules.

(a) State two other typical properties of copper and its compounds that are different from those of non-transition metals.

Colour ✗

It shows variable oxidation states. ✓

[2 marks] 1/2

(b) Explain how a water molecule behaves as a ligand.

It can bond with the empty d orbitals of the transition metal ion. ✗

[2 marks] 0/2

(c) Aqueous copper(II) sulphate contains the complex ion $[Cu(H_2O)_6]^{2+}$. When an excess of aqueous ammonia is added to aqueous copper(II) sulphate, ligand substitution takes place.

(i) What would you see?

The solution goes a deep blue-violet colour. ✓

(ii) Write an equation for this ligand substitution.

$[Cu(H_2O)_6]^{2+} + 4NH_3 \rightarrow [Cu(NH_3)_4]^{2+} + 6H_2O$ ✗ ✓

[3 marks] 2/3

[Total 7 marks]

3/7

2 Iron is the most useful of all metals. Its ions have incompletely filled d-orbitals, and so it has the typical properties of a transition metal.

(a) When a solution of sodium hydroxide is added to a solution containing **hydrated** iron(II) ions, a green precipitate is formed. This precipitate slowly turns red-brown on exposure to air.

Write equations for these reactions and explain the processes taking place.

$Fe^{2+}(aq)$ ✗ $+ 2OH^-(aq) \rightarrow Fe(OH)_2(s)$

$2Fe(OH)_2 + O_2 + H_2O \rightarrow 2Fe(OH)_3$ ✓ ✗

The first reaction is precipitation, ✗ and the second is oxidation. ✓

[5 marks] 2/5

(b) When a solution of potassium cyanide, KCN(aq), is added to a solution containing hydrated iron(II) ions, the complex $[Fe(CN)_6]^{4-}$ ion is formed.

(i) What type of reaction is occurring?

It is a ligand exchange reaction. ✓

(ii) What is the bonding in the complex ion?

The bonding is dative covalent between the ligand and the Fe^{2+} ion ✓ ∧

(iii) Draw and name the shape that you would predict for this complex ion.

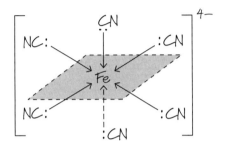

The shape is octahedral ✓

[5 marks] ④/5

⑥/10 [Total 10 marks]

How to score full marks

1 (a) You must make it clear that it is the **ions** that are coloured (only when complexed), and **not the element** itself. The student could have chosen any two of: '(i) form coloured ions; (ii) show variable oxidation states in their compounds; (iii) their ions are paramagnetic; (iv) the elements and their compounds are often good catalysts; (v) the elements have a higher density and melting point than s or p block metals.'

(b) You must **explain the nature of the bonding**, which is: 'the lone pair of electrons on the ligand forms a dative covalent (co-ordinate) bond with the transition metal ion.'

(c) The student correctly showed the four NH_3 molecules as ligands, but the copper/ammonia complex still has **two molecules of water** in its co-ordination sphere. The correct equation is:

$$[Cu(H_2O)_6]^{2+} + 4NH_3 \rightarrow [Cu(NH_3)_4(H_2O)_2]^{2+} + 4H_2O$$

2 (a) The first equation may look correct, but the student should have used the **proper formula for the hydrated ion**. The second equation gained one mark for the correct species, but lost the second because it did not **balance**. The two equations should be:

$$[Fe(H_2O)_6]^{2+} + 2OH^- \rightarrow Fe(OH)_2 + 6H_2O$$

$$4Fe(OH)_2 + O_2 + 2H_2O \rightarrow 4Fe(OH)_3$$

The student made a common error by merely **restating the question** with the word 'precipitation'. The correct description of this type of reaction is 'deprotonation'.

(b) (ii) The student forgot to mention the **covalent (triple) bond** within the cyanide ion.

Don't make these mistakes...

Don't confuse *d*-block with transition metals. Scandium and zinc are not transition metals as their ions do not have partially filled *d*-orbitals.

Remember, when a (period 4) transition metal forms an ion it always **loses its 4s electrons**. It may also lose some of its 3*d* electrons.

Don't forget that hydrated ions are all complexes with water. Always write their formulae as the $[M(H_2O)_6]^{x+}$ ion.

When a hydrated transition metal ion is reacted with an alkali such as sodium hydroxide, it is **deprotonated**, and a precipitate of the metal hydroxide is formed, but if ammonia, or other ligands, are added, **ligand substitution** may then take place. **Don't confuse these two types of reaction.**

Don't miss important clues for helping you understand what type of reaction is happening. The **colour** of an ion **changes** if either the **oxidation number** changes or **ligand substitution** takes place.

$Cr(OH)_3$ is **amphoteric** (as is the *d*-block $Zn(OH)_2$ and the *p*-block $Al(OH)_3$) and so **reacts with acids** – to form $[Cr(H_2O)_6]^{3+}$, and with **alkalis** – to form $[Cr(OH)_6]^{3-}$.

When you draw the shape of a complex ion, it is always a good idea to **give the name of the shape** in case your drawing is not very clear.

Key points to remember

- In a ***d*-block** element the highest energy electron is in a ***d*-orbital**. In a **transition metal** at least of its ions has a **partly filled *d* shell**.

- The number of d electrons **increases** one at a time from scandium to zinc, **except** that chromium is [Ar] $3d^5\ 4s^1$ not [Ar] $3d^4\ 4s^2$, and copper is [Ar] $3d^{10}\ 4s^1$ not [Ar] $3d^9\ 4s^2$.

- Complex ions result when a lone pair of electrons in the ligand forms a dative covalent bond with the transition metal ion. Usually there are six ligands per ion, but sometimes only four, as in $[CuCl_4]^{2-}$.

- When there are **six ligands**, the ion is **octahedral**.

- The **colour** is caused by the ion **absorbing** a photon, which promotes a *d* electron from one *d* orbital to another *d* orbital. The *d* orbitals are split into two levels by the bonded ligands.

- **Variable oxidation state** is the result of a steady increase in successive ionisation energies, unlike the *s* and *p* block elements, where there is a sudden increase when electrons from an inner shell are removed.

- The hydrated ions are **deprotonated** by bases and form coloured **precipitates** of metal hydroxides.

- The hydrated ions undergo **ligand exchange** with suitable ligands. Typical ligands are NH_3, Cl^-, CN^-, bidentate ligands such as diethanoate (oxalate), $C_2O_4^{2-}$, and polydentate ligands such as EDTA.

- The **reduction potential** ($E^{\ominus}$) of a **complexed** ion is **different** from that of the hydrated ion.

Typical transition metals

- Form **complex ions** that are **coloured**.

- Have **several oxidation states** in their compounds.

- Are good **catalysts** (as are their compounds).

- Form **paramagnetic** ions.

- Are **denser** and have **higher melting points** than *s* and *p* block metals.

Questions to try

Q1

(a) (i) Write the electron configurations for Cu and for Cu$^+$.

....

[1 mark]

(ii) In what way is the electronic structure of copper unusual in terms of the general trend across the first transition series?

....

[1 mark]

(iii) Explain why the copper(I) ion is not coloured.

....

....

[2 marks]

(b) Use the electrode potential data concerning copper to answer the questions that follow.

$Cu^{2+}(aq) + e^- \rightarrow Cu+(aq)$ $\qquad\qquad\qquad\qquad\qquad\qquad$ $E = +0.15V$

$Cu^+(aq) + e^- \rightarrow Cu(s)$ $\qquad\qquad\qquad\qquad\qquad\qquad\quad$ $E = +0.52V$

Suggest what would happen if a sample of copper(I) sulphate, Cu_2SO_4, was added to water. State in general terms the nature of the process which is occurring and state what you would see when copper(I) sulphate was added to water.

....

....

....

[7 marks]

(c) Copper(II) sulphate solution contains the complex ion $[Cu(H_2O)_6]^{2+}$.

(i) Draw the 'electrons in boxes' diagram for this complex ion, distinguishing clearly the copper electrons from the ligand electrons.

[2 marks]

(ii) State what you would see if a solution of aqueous ammonia was added dropwise to a solution of this ion. Show by means of equations how this reaction proceeds and state the type of reaction occurring at each stage.

....

....

....

....

[5 marks]

[Total 18 marks]

Examiner's hints for question 2
(a) Select your transition element after you have written down the four properties. Make sure that you can give specific examples.
(b) The main clues here are the results of tests (ii) and (iii). Remember that you are expected to identify both the cation and the anion in A. Remember that the colour alters if either the ligand or the oxidation state is changed.

Q2

(a) State **four** typical properties shown by transition metal elements or their compounds, illustrating your answer with any **one** transition metal.

....

....

....

....

[8 marks]

(b) A green solid **A** was dissolved in a little distilled water.

- When this solution was diluted, it turned into a blue solution **B**.

- Some dilute nitric acid and silver nitrate solution were added to one portion of this diluted solution and a white precipitate **X** was obtained.

- Some dilute sodium hydroxide was added to another portion of the diluted solution, and a blue precipitate **Y** was obtained.

- Excess ammonia solution was added to a third portion of the diluted solution, and a deep blue-violet solution **Z** was obtained.

(i) Identify the solid **A**, the solutions **B** and **Z** and the precipitates **X** and **Y**.

....

....

....

....

....

[5 marks]

(ii) Write the ionic equation for the formation of the white precipitate **X**.

....

[2 marks]

The answers to these questions are given on pages 92–93.   [Total 15 marks]

10 How to revise for synoptic questions

Synoptic questions truly test your ability as a chemist. To succeed you have to revise more than one unit, and for some papers you must revise the whole syllabus. You must be prepared to **link your understanding of one topic with another**.

In chemistry the links are the compounds themselves, and a useful way of preparing yourself is to draw up and study 'mind maps' or 'spidergrams'. I have made up four of these. Follow the threads and check that you understand the chemical principles touched on in the spidergrams. This will help you to consolidate your understanding.

When you have studied the spidergrams below, try to make up some of your own. You might start by drawing them for water, sodium hydroxide, sulphuric acid and then chlorine. Link topics such as structure and bonding, acidity, organic reactions, redox reactions, equilibrium, kinetics and manufacture through each substance.

In organic chemistry look for links with physical chemistry (enthalpy, equilibrium, kinetics) and with inorganic substances that they react with. A good choice for an organic spidergram would be 2-bromopropane.

Spidergram for ammonia:

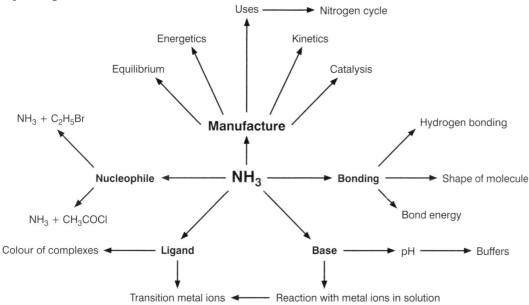

Look at the spidergram and check your understanding of the following topics.

- **The manufacture of ammonia:** you must know the conditions of temperature, pressure and the name of the catalyst used. You must understand the effect of temperature and pressure on yield (equilibrium) and the rate of the reaction (kinetics). You must know that a catalyst does not alter the position of equilibrium, but that it allows the reaction to take place at an economic rate at a lower temperature.

- You need to know some **uses of ammonia**, and understand the effect on the environment of an overuse of nitrogenous fertilisers.

- You must be able to **draw the electronic structure of ammonia**, and hence be able to **explain its shape**, and **why it can hydrogen bond** with other molecules of ammonia and with water. You must be prepared to do **bond energy calculations**.

- You should know the **definition of a nucleophile**, and hence why ammonia is a nucleophile. You must understand that certain organic reactions (with halogenoalkanes and with acid chlorides) and inorganic reactions (ligand exchange) are caused by ammonia's ability to act as a nucleophile. You need to be able to **explain the change of colour** that happens when ligand exchange takes place.

- **Ammonia is also a base.** You must be able to explain this in terms of accepting a proton, and forming OH^- ions in solution and hence the action of aqueous ammonia with acids and with metal ions. You must understand the concept of a conjugate acid/base pair, and you must be able to explain the action of an alkaline buffer solution.

Spidergram for magnesium sulphate:

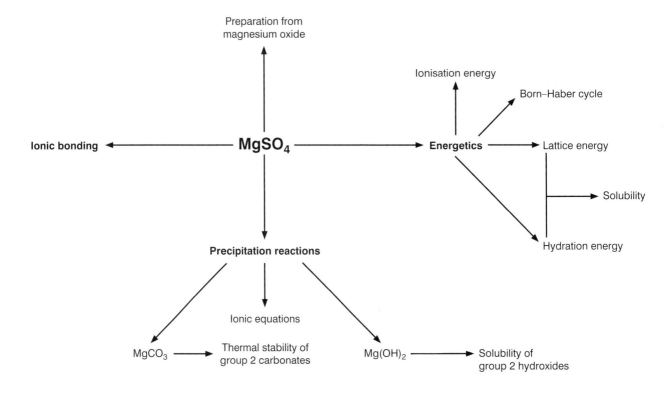

From the spidergram, check that you know and understand the following facts and concepts:

- **Ionic bonding:** you must be able to explain this in terms of attraction between ions of opposite charge and of electron transfer. Ionic bonding links to energetics. Understand the Born-Haber cycle, and terms such as first and subsequent ionisation energies, electron affinity, the enthalpy of atomisation and, in particular, lattice energy.

- You need to appreciate that the **energy change** in dissolving an ionic solid can be regarded as the sum of the energy changes for splitting up the lattice and hydrating the gaseous ions.

- Magnesium sulphate undergoes **precipitation reactions**. In particular, you need to know that hydroxides can be precipitated by adding a solution containing hydroxide ions from a soluble hydroxide such as sodium hydroxide. Also carbonates can be prepared by adding a solution containing carbonate ions, $CO_3^{2-}(aq)$, from a soluble carbonate such as ammonium carbonate.

- Heating can **decompose** group 2 carbonates, as with all carbonates except those of the group 1 metals. The stability to heat increases down the group as the polarising power of the cation decreases. This is because ions of high polarising power can more easily pull off an oxide ion from the carbonate ion.

- The **solubility** of the group 2 hydroxides **increases down the group**. The entropy changes are much the same for dissolving any of them, but the value of $\Delta H_{solution}$ becomes less endothermic going from magnesium hydroxide to barium hydroxide because the lattice enthalpy decreases more than the hydration enthalpy of the cation decreases as the positive and negative ions are of a similar size.

- You need to know how a pure sample of magnesium sulphate can be prepared. This includes an understanding of the process of recrystallisation.

Spidergram for vanadium:

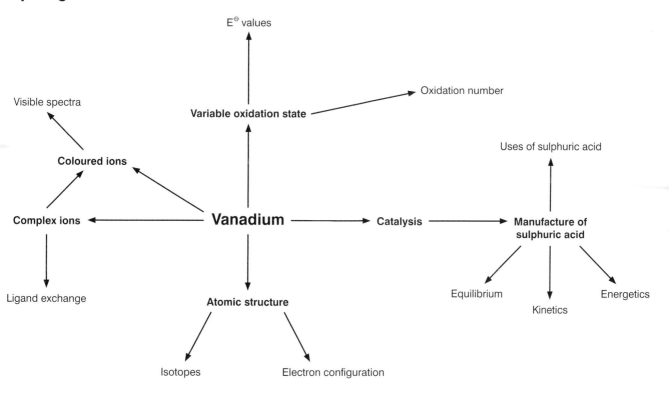

The chemistry of vanadium is useful in helping you to revise a number of different topics.

- You must be able to explain how **isotopes** differ from each other, and must understand the difference between relative isotopic and atomic masses.

- You need to understand the **aufbau principle** and **Hund's rule** and be able to apply them to writing the electronic structures of elements up to number 38.

- You must know the major properties associated with **transition metals** and their compounds.

- Check that you understand the bonding in **complex ions** and their shape.

- Make sure you understand why **complex ions are normally coloured** and why complex ions of scandium(III), titanium(IV), copper(I) and zinc(II) are not coloured. You must know why a change of ligand or oxidation state alters the colour of the complex.

- You need to be able to explain, in terms of energy, why **transition metals are stable** in a variety of oxidation states. You should be able to calculate $E^{\ominus}$ values for redox reactions, and hence comment on their feasibility.
  ($E^{\ominus}$ calculations are not required by OCR.)

- You must appreciate that many transition metals and their compounds are used as **catalysts**, and the difference between homogeneous and heterogeneous catalysts. You should know examples such as iron in the Haber process, nickel or platinum in the hydrogenation of alkenes. You need to be able to calculate $\Delta H_{reaction}$ for the critical step in the manufacture of sulphuric acid, and comment, in terms of equilibrium and kinetics, on the economics of the process. You must know some uses of the acid.

Spidergram for carboxylic acids:

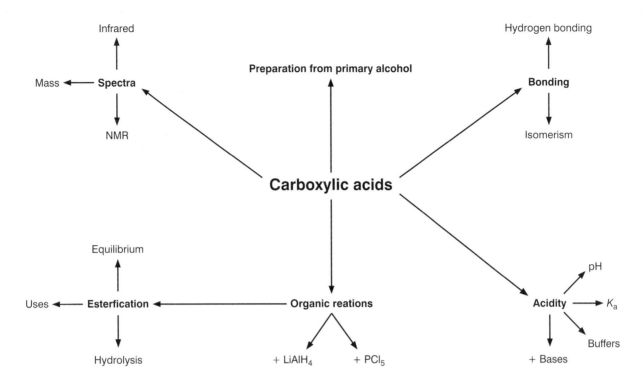

Carboxylic acids link organic reactions with physical chemistry topics such as ionic equilibria and inter-molecular forces. In addition they provide examples of all types of spectra.

- **Organic reactions** of carboxylic acids, such as ethanoic acid and benzoic acid, and their preparation: check that you can write equations, know the conditions and can write the structural formulae of the products of:

  1 reactions with lithium aluminium hydride and with phosphorus pentachloride

  2 esterification. This reaction links the calculation of $\Delta H_{reaction}$ to bond energies and the effect of temperature on equilibrium.

- **Ionic equilibria:** Carboxylic acids are weak acids. You must be able to:

  1 understand the concept of K_a and hence calculate the pH of an acid's solution

  2 explain how a buffer solution maintains a nearly constant pH

  3 calculate the pH of a buffer solution.

- **Bonding:** You must be able to write the structural formulae of a carboxylic acid and explain the C=O bond in terms of σ and π bonding. You need to understand that carboxylic acids can form hydrogen bonds between molecules, and that this leads to the formation of dimers. They also hydrogen bond with water, which helps to make them soluble. Acids such as C_3H_7COOH show structural isomerism, amino acids show optical isomerism and unsaturated acids such as HOOCCH=CHCOOH show geometric isomerism.

- **Spectra:** Carboxylic acids exhibit typical C=O and O–H infrared absorption frequencies. They also absorb UV light because of their carbon/oxygen π bond. You must be able to assign structures to mass spectrum fragments, and explain the chemical shift and spin coupling in NMR spectra.

The different types of synoptic questions

Synoptic questions will be found in Unit 6 papers in all exam boards and in some questions in Unit 5 papers in all boards except Nuffield.

The questions are often open-ended and require you to make links between different topics. They also **test your ability in synthesis of knowledge and understanding** rather than factual recall. The syllabus defines this as:

'students should be able to bring together knowledge, principles and concepts from different areas of chemistry, and apply them in a particular context, expressing ideas clearly and logically'.

You will frequently see the words 'deduce' or 'predict' and less often the words 'what happens when' or 'state'. You will find that calculations are not broken down into separate parts, as you are expected to be able to see your way through a multi-step calculation. AQA uses objective tests in its main synoptic paper. Here you must be able to move mentally from topic to topic as you progress through the paper.

There are several types of synoptic question:

Question type A:	OCR and AQA ask very open-ended essay-type questions. (See pages 71 and 78.)
Question type B:	You are faced with an entirely new situation, in some if not all, of the question. However, you will be given enough information to solve the problem. This type usually involves a calculation. (See page 74.) This type is used by all boards.
Question type C:	AQA uses objective testing (multiple choice and multiple completion) in Unit 6, which is its main synoptic paper. (See pages 76 and 78.)
Question type D:	This type of question is about a complex compound or one that is new to you. You are not expected to know the reactions of this compound but should be able to deduce them from your knowledge of compounds with similar functional groups. (See page 79.)
Question type E:	Some synoptic questions, especially those in Edexcel papers, link several different areas with a common theme or a common compound. (See page 80.)
Question type F:	You could be asked to plan an experiment that would be similar to one you have done in the laboratory, but you will have to adapt a known technique to a new situation. (See page 81.)

How to score full marks

Question type A

Write an essay on the chemistry of ammonia.

In your answer you should include a discussion of the bonding in, and the structure of the molecule, consider its industrial preparation and you should bring together its reactions in both inorganic and organic chemistry.

[15 marks]

How to tackle this question

You must plan these open-ended questions. First read the question carefully and write down, on rough paper, **the headings** and your **estimate of the marks** that will be awarded under each heading. In this question there are four headings, so you must try to make **at least four scoring points** under each heading. Then add, briefly, what scoring points you will make. Your notes should be similar to these:

> - The bonding and structure of ammonia: 4 marks
>   Electronic structure of nitrogen, type of bonding in ammonia, shape.
> - Its industrial preparation: 4 marks
>   Equation, conditions (P,T and catalyst), reasons for conditions.
> - Its inorganic reactions: 3 or 4 marks
>   Deprotonation of water and aqua cations, reaction with acids (inorganic and organic)
> - Its organic reactions: 3 or 4 marks
>   Nucleophile (+ explanation of meaning), reaction with halogenoalkanes and with acid chlorides.

Now you are ready to write your answers. You should do so under the **four headings** – and within those headings it is a good idea to make **bulleted points**. Extra marks will be awarded for clarity of expression and good English, so within the bulleted points write **proper sentences**. You must make at least **4 points per heading**.

An answer that scores full marks will include all or most of the following points:

> Bonding and structure
> - N has an electronic structure of $1s^2$, $2s^2$ $2p^3$ ✓ and so forms 3 covalent bonds with hydrogen. ✓
> - The electronic structure of the ammonia molecule is
>
> $$H \overset{\times}{\underset{\times}{\bullet}} N \overset{\bullet\bullet}{\underset{\bullet}{\times}} H \quad ✓$$
>
> - Its shape is pyramidal, ✓ with a HNH bond angle of $104\frac{1}{2}°$. ✓ This is because the molecule has 3 bond pairs and 1 lone pair ✓ of electrons which repel each other and get as far apart as possible.
> Score: maximum of 4

Manufacture

- Ammonia is manufactured by the Haber process
  $$N_2 + 3H_2 \rightleftharpoons 2NH_3 \checkmark$$
- The reaction needs a high pressure of about 250 atm to drive the equilibrium to the right ✓
- and a temperature of about 400°C. which is a compromise between high temperature, which gives a fast reaction but a low yield, ✓ and low temperature, which gives a high yield but too slow a reaction. ✓
- A catalyst of iron ✓ is used so that the reaction can proceed quickly and at a moderate temperature. ✓

Score: maximum of 4

Inorganic reactions

- It will deprotonate water and so its solution is alkaline:
  $$NH_3 + H_2O \rightleftharpoons NH_4^+ + OH^- \checkmark$$
- It removes protons from aqua cations and so precipitates some metal hydroxides:
  $$Al(H_2O)_6^{3+} + 3NH_3 \rightarrow Al(OH)_3 + 3NH_4^+ + 6H_2O \checkmark$$
- It forms salts with inorganic acids:
  $$NH_3 + HCl \rightarrow NH_4^+ Cl^{-s} \checkmark$$
- and with organic acids:
  $$NH_3 + CH_3COOH \rightarrow CH_3COO^-NH_4^+ \checkmark$$
- It can form ligands with transition metal ions: ✓
  $$4NH_3 + [Cu(H_2O)_6]^{2+} \rightarrow [Cu(NH_3)_4(H_2)_2]^{2+} + 4H_2O \checkmark$$

Score: maximum of 4

Organic reactions

- It is a nucleophile, ✓ as it can donate a lone pair of electrons and form a bond with electron-deficient atoms.
- Thus it causes nucleophilic substitution with halogenoalkanes:
  $$2NH_3 + C_2H_5Br \rightarrow C_2H_5NH_2 + NH_4Br \checkmark$$

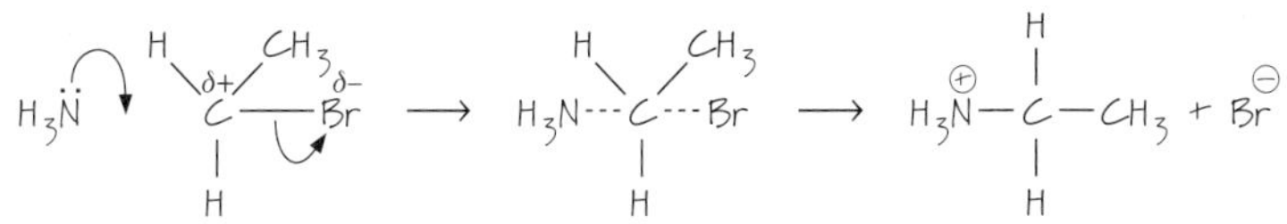

- It also reacts as a nucleophile with acid chlorides:
  $$2NH_3 + CH_3COCl \rightarrow CH_3CONH_2 + NH_4Cl \checkmark$$

Score: maximum of 4

Maximum score for the question: 15

Note that there are 6 possible scoring points in each section. **Always try to give more scoring points than the minimum required by the examiner.**

Question type B

A solution of cobalt(II) chloride was reacted with ammonia and ammonium chloride while a current of air was blown through the mixture. A red compound, **X**, was produced which contained a complex ion of cobalt. The compound had the following composition.

Element	Co	N	H	Cl
% by mass	23.6	27.9	6.0	42.3

(a)   Use the data to calculate the empirical formula of the compound **X**. [3 marks]

(b)   In the reaction in which **X** is formed from cobalt(II) chloride, explain the role of:

  (i) the air

  (ii) the ammonium chloride.

[3 marks]

(c)   A solution of cobalt(II) chloride reacts with concentrated hydrochloric acid to form a stable complex ion. A solution of calcium chloride does not form a corresponding complex. Use your knowledge of atomic structure to explain this difference, and suggest other differences you would expect between the chemistry of cobalt and calcium.

[4 marks]

How to tackle this question

(a)   Set your calculation out clearly, **showing what you are doing at each step**.

	%	% ÷ Ar (= moles) ✓	÷ smallest (= simplest mole ratio) ✓	
Co	23.6	÷ 58.9 = 0.40	0.40 ÷ 0.40 = 1	
N	27.9	÷ 14.0 = 1.99	1.99 ÷ 0.40 = 5.0	Empirical formula is $CoN_5H_{15}Cl_3$ ✓
H	6.0	÷ 1.0 = 6.0	6.0 ÷ 0.40 = 15.0	
Cl	42.3	÷ 35.5 = 1.19	1.19 ÷ 0.40 = 3.0	

(b) **(i)** Cobalt is a transition metal and hence shows variable valency. Thus cobalt(II) can be oxidised, and so the function of the air is that

> air is an oxidising agent ✓ and oxidises the Co^{2+} ions to a higher oxidation state. ✓

(ii) The mixture of a weak base (ammonia) and its salt (ammonium chloride) should make you think of buffer solutions. Your answer should be:

> The function of the ammonium chloride is that it, together with the ammonia, forms a buffer solution. ✓ Also the original compound has 2 Cl atoms to 1 Co atom, whereas the compound X has 3 Cl atoms to each Co ✓ atom, and so the ammonium chloride also acts as a source of Cl^- ions.

(c) The clue here is that cobalt is a transition metal, whereas calcium is not. You must explain why cobalt can form a complex, and why calcium cannot. Then you list, but do not explain, three other properties typically found in transition metals but not in *s* block metals. Your answer should be similar to this:

> Co^{2+} can use its 3d and 4p orbitals ✓ to accept pairs of electrons in complex ion formation from both NH_3 and Cl^-, ✓ whereas Ca^{2+} ions cannot as calcium is not a transition metal. ✓
>
> Cobalt and its compounds can act as catalysts, but calcium does not, ✓ its complex ions are coloured, but calcium ions are not ✓ and cobalt compounds exist in more than one oxidation state, whereas all calcium compounds are 2+ only. ✓

How to score full marks

Question type C (for AQA only)

Unit 6 for AQA candidates is entirely an objective test paper with a number of multiple choice (where only one answer is correct) and a larger number of multiple completion questions. The tactic for scoring high marks is different for the two. In each case you will be given a separate answer grid for your answers, so you can write your thoughts on the exam paper.

Multiple choice

- Look at all four answers on the question paper. Pencil a tick by the answers you think may be right and a cross by those you think are wrong. If you have two ticks, look carefully and decide which of the two answers you are more sure of. Then mark the correct letter on your answer sheet.

- For questions with a negative (which does **not ...**) put a cross by each answer you think fits. If you have two crosses, look carefully and decide which you are more sure of.

- Don't stop checking all the offered answers just because you think that you have found the right one.

- If the question is a calculation, do the calculation and then see which answer fits yours. If none does, try to find your error.

- If you cannot answer the question, don't waste time. Put a large circle around the question's number. Leave the question and come back at the end, if you have time.

- **Never leave a question unanswered.** At the end go back and **guess** a letter for any unanswered question. Marks are not deducted for wrong answers and you might guess right.

Some examples are:

> **3** Which of the following pairs will give a racemic mixture?
>
> **A** $CH_3CH=CH_2 + HBr$ ✓
>
> **B** $CH_3CH_2CHO + HCN$ ✓
>
> **C** $CH_3COCH_3 + NaBH_4$ ✗
>
> **D** $CH_3CH_2COCl + CH_3NH_2$ ✗

Working:

*A and B can't both be right. Aldehydes add with HCN to give a compound with 4 different groups on a carbon atom. The attack can come from either side so a racemic mixture will be produced. The addition to asymmetric alkenes goes via Markovnikov's rule, which means that the product in A will be $CH_3CHBrCH_3$, which is not a chiral molecule, and so a single isomer will be produced. So **B** is correct.*

> **4** Which one of the following reactions does **not** produce ethanol?
>
> **A** CH_3CH_2Br warmed with aqueous sodium hydroxide
>
> **B** CH_3COOCH_3 warmed with aqueous sodium hydroxide ✗
>
> **C** CH_3CHO vapour passed with hydrogen over hot platinum ✗
>
> **D** $CH_2=CH_2$ passed with steam over phosphoric acid at high pressure

Working:

*You may have thought that aldehydes cannot be reduced catalytically by hydrogen, but the ester is a **methyl** ester and so cannot possibly form ethanol when hydrolysed. So ethanol won't form in **B** and so **B** is the answer.*

Multiple completion

- The rubric for these is: decide which of the responses to the question is/are correct and select **A**, **B**, **C** or **D** as follows:

A	B	C	D
(i), (ii) and (iii) only	(i) and (iii) only	(ii) and (iv) only	(iv) only

- Again, use your exam paper to think out your answers before filling in the answer grid.

- For these questions, go through all four responses and put a tick if you think that the answer is correct and a cross if you think it is wrong.

- Now check that your set of ticks matches up with allowed combinations.

- If not, see which one you are certain of and give it a double tick. Now see what possible combinations go with this.

- For example, if you are certain of (i), then you will know from the possibilities allowed that (iv) cannot be right. So the answer must be **A** or **B**. You now check (ii). If you cannot decide whether (ii) is correct guess **A** or **B**.

- If you are certain that (ii) is right, the answer must be **A** or **C**. If (i) is wrong, then the answer must be **C**.

Some examples (all taken from the AQA unit 6 specimen paper) are:

5 Ionic bonding is found in

   (i) NH_4Cl ✔

   (ii) $NaBH_4$ ✔ ✔

   (iii) CH_3COOK ✔ ✔

   (iv) $BeCl_2$ ✔

Working:

*There is no response that (i), (ii), (iii) and (iv) are all correct. So think again. (ii) and (iii) **must** be correct as all sodium and potassium compounds are ionic. The only combination containing (ii) and (iii) is **A**, so (i) must also be ionic and so give **A** as your answer.*

6 All the atoms lie in a single plane in molecules of

   (i) BCl_3 ✔

   (ii) HCHO ✔

   (iii) $C_2H_2Br_2$ ✗

   (iv) NH_3 ✗

Working:

*There is no response (i) and (ii) only. If you are sure that both are planar molecules, the only response with both (i) and (ii) is **A**. That means that $C_2H_2Br_2$ must also be planar. (It **is** planar as is ethene.)*

7 Brønsted–Lowry acid–base reactions include

   (i) $OH^- + CH_3Cl \rightarrow CH_3OH + Cl^-$ ✔

   (ii) $NH_3 + HCl \rightarrow NH_4^+ + Cl^-$ ✔ ✔

   (iii) $KF + PF_5 \rightarrow K^+ + PF_6^-$ ✗

   (iv) $H_2O + H_2O \rightarrow H_3O^+ + OH^-$ ✔

Working:

*There is no response (i), (ii) and (iv). If you are sure that (ii) is an acid–base reaction, then you have to decide whether (i) or (iv) is correct. Looking carefully at (i) you can see that the OH^- ion is **not** acting as a base as it does **not** accept an H^+ ion to form water. On checking you should realise that (iv) is a proton exchange reaction. Thus the correct statements are (ii) and (iv) and so you should mark **C** as your answer.*

8 Which of the following statements about a catalyst is/are true?

(i) It speeds up the forward and slows down the reverse reaction. ✗

(ii) It increases the proportion of molecules with higher energies. ✗

(iii) A homogeneous catalyst usually acts in the solid state. ✓

(iv) It does not alter the value of the equilibrium constant. ✓✓

Working:

There is no response (iii) and (iv). You should be certain of (iv). Thus either all the others are wrong or (ii) is correct. Catalysts do not increase the temperature of gases. Looking at (iii) again, you ought to realise that homogeneous means 'in the same phase', and solid catalysts are usually used in gas reactions, so (iii) is wrong. This confirms (iv) as the only correct statement and so you should mark **D** *as your answer.*

Questions to try

Question type A

Examiner's hints *for question 1*
- This is a very open-ended question, and so your answer must be planned.
- You should start by sketching out the sort of points and links that you will make. There are 19 chemical scoring points and 3 for quality of language.

  Bonding ⟶ forces between particles ⟶ boiling points (about 7 scoring points)

  Bonding ⟶ reaction (if any) with water ⟶ conductivity (about 7 scoring points)
                                      └⟶ pH (about 7 scoring points)

- Remember to give equations for reactions, and to use quantitative data if possible – use $[H^+]$ values to calculate pH and K_a for any weak acids.

Q1

Using knowledge, principles and concepts from different areas of chemistry, explain and interpret, as fully as you can, the data given in the table below. In order to gain full credit, you will need to consider each type of information separately and also to link this information together. [22 marks]

Compound	Boiling point/K	Properties of a 0.1 mol dm⁻¹ solution	
		Electrical conductivity	$[H^+]$/mol dm⁻³
NaCl	1686	Good	1.0×10^{-7}
CH_3COOH	391	Slight	1.3×10^{-3}
CH_3CH_2OH	352	Poor	1.0×10^{-7}
$AlCl_3$	451	Good	3.0×10^{-1}

Question type D

Examiner's hints *for question 2*

Don't get put off by the size of the molecule. The question is simply about the chemistry of alkenes and primary alcohols.

(a) (i) Before you answer this, ask yourself two questions. Can bombykol form hydrogen bonds with water? Does it contain a long non-polar 'tail'? Then discuss whether it will be soluble, slightly soluble or insoluble in water.

(ii) Don't forget that there are two C=C groups in the molecule, and thus it may show geometric (*cis/trans*) isomerism around both double bonds.

(iii) What adds to one double bond will add to the other. Bombykol also contains an OH group, and so you should choose reaction of both types of functional group. Equations must balance, reaction schemes need not. In both cases you should state special conditions, such as catalyst.

(b) You probably haven't heard of ozonolysis, but don't panic as the reaction is explained. Remember that the initial products are aldehydes, and that the reaction is performed under oxidising conditions.

Q2

The structure of the silkworm moth sex attractant, bombykol, is

$H_3C–(CH_2)_2–CH=CH–CH=CH–(CH_2)_8–CH_2OH$

(a) Predict some of the properties you would expect for bombykol. You should comment on:

(i) the likely solubility of bombykol in water;

(ii) the number of possible geometric isomers

(iii) its likely reaction with four reactants of your choice.

Write equations or reaction schemes for the reactions you choose, showing the structures of the organic products clearly.

[8 marks]

(b) When bombykol is treated is treated with ozone in the presence of water, the molecule splits into fragments wherever there is a C=C double bond.

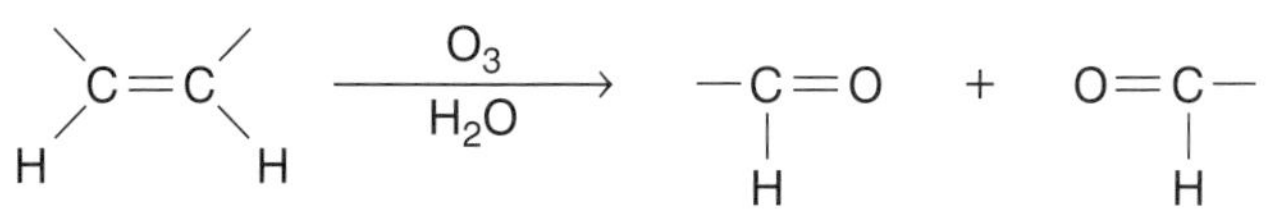

The aldehyde groups are converted to carboxylic acids in oxidising conditions Predict the oxidation products formed when bombykol is reacted with ozone in oxidising conditions.

[2 marks]

Question type E

Examiner's hints *for question 3*

The link that connects all the parts of this question is that the substance is a base.
(a) Use the % mass data to calculate the mole ratio of the elements and hence the empirical formula.
(b) All titration curves, except those of a weak acid with a weak base, have a vertical portion at the equivalence (end) point. Having found the end point, you can calculate the concentration of the acid in mol dm^{-3}, and hence you can evaluate the relative molecular mass as you are given the concentration in g dm^{-3}.
(c) (ii) Remember that **X** is a primary amine and that it does not react with bromine.
(d) (i) Give the names or formulae for the reagents and what you observe for all tests.
(ii) Compare the *m/e* values with the molar mass, and hence see what fragments have been lost to give peaks at 57 and 58. Then say whether both or just one isomer would produce these.

Q3

Compound **X** has a primary amine group and contains 49.3% carbon, 19.2% nitrogen, 9.6% hydrogen and 21.9% oxygen by mass. It does **not** react with bromine water.

When a solution containing 6.07 g dm^{-3} of **X** was steadily added to 20.0 cm^3 of a 0.104 mol dm^{-3} solution of hydrochloric acid, the following pH graph was obtained:

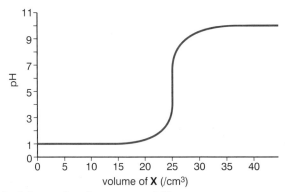

(a) Calculate the empirical formula of compound **X**. [2 marks]

(b) Use the graph to determine the end point, and hence calculate the molar mass of compound **X**. [5 marks]

(c) (i) Use your result from part **(a)** and **(b)** to calculate the molecular formula of **X**. [1 mark]

(ii) Suggest two structures for **X** that are consistent with the information given above. [2 marks]

(d) (i) Describe **a series of tests** you could perform in order to identify the functional group other than the amine group in **X** and which also distinguishes **X** from the other isomer you suggested in **(c) (ii)**. [4 marks]

(ii) The mass spectrum of **X** includes major peaks at *m/e* values of 57 and 58. Show if this data is consistent with only one of the structures for **X**. [3 marks]

Question type F

Examiner's hints *for question 4*

(a) You must first work out which substance is in excess and which is the limiting reagent.

(b) The answer should be in three parts:

- A labelled diagram of the apparatus that you would use.
- Suitable quantities of all chemicals.
- The details (preferably in a numbered list of instructions) of how you would carry out the experiment at each temperature.

Q4

When solutions of ammonium chloride, NH_4Cl, and sodium nitrite, $NaNO_2$, are mixed, nitrogen gas is steadily produced. The reaction is complete within about two minutes.

$$NH_4Cl(aq) + NaNO_2(aq) \rightarrow 2H_2O(l) + N_2(g) + NaCl(aq)$$

(a) Calculate the volume of nitrogen gas formed when 100 cm^3 of 0.100 mol dm^{-3} ammonium chloride solution is mixed with 80 cm^3 of 0.111 mol dm^{-3} sodium nitrite solution. You may assume that one mole of gas occupies 24.0 dm^3 under room conditions of temperature and pressure. [5 marks]

(b) Describe, giving full details of quantities, apparatus and procedure, how you would carry out experiments to show how much faster the reaction proceeds at 35°C than at room temperature. Use your answer to part **(a)** to help you choose a suitable scale of apparatus and amount of each solution to use. [10 marks]

Question type C (for AQA candidates only)

Examiner's hints for questions 5–11

- See 'How to score full marks' on page xxx.

Multiple choice questions

(For these questions, select the best response)

Q5 Which substance is acting as an oxidising agent in the reaction:

$$10HCl + 2MnO_2 + 3KBiO_3 \rightarrow 2KMnO_4 + 3BiCl_3 + 5H_2O + KCl$$

A $KMnO_4$

B MnO_2

C $KBiO_3$

D HCl

Q6 Which one is **not** hydrogen bonded in the liquid state?

A CH_3CONH_2

B CH_3NH_2

C $(CH_3)_2NH$

D $(CH_3)_3N$

 Q7 The electrode potential of the cell, for which the overall equation is shown below, was found to be 1.26V, whereas the standard electrode potential for this cell is 1.24V

$$Fe(s) + 2Ag^+(aq) \rightleftharpoons Fe2^+(aq) + 2Ag(s)$$

Which of the following could account for the difference in potential?

A The concentration of the Fe^{2+} ions was 1.0 mol dm^{-3} and the concentration of the Ag^+ ions was 2.0 mol dm^{-3}.

B The iron electrode was smaller than the silver electrode.

C The volumes of the solutions in each half-cell were different.

D The salt bridge was omitted.

Multiple completion questions

Decide which of the responses to the question is/are correct and select **A**, **B**, **C** or **D** as follows

A	B	C	D
(i), (ii) and (iii) only	(i) and (iii) only	(ii) and (iv) only	(iv) only

Q8 Which of the following reactions has a negative value of $\Delta S^{\ominus}$?

(i) $Fe^{2+}(aq) + 2OH^-(aq) \rightarrow Fe(OH)_2(s)$

(ii) $CaCO_3(s) \rightarrow CaO(s) + CO_2(g)$

(iii) $H_2(g) + C_2H_4(g) \rightarrow C_2H_6(g)$.

(iv) $6CO_2(aq) + 6H_2O(l) \rightarrow C_6H_{12}O_6(aq) + 6O_2(g)$

Q9 Which of the following is/are always correct about the effect of increasing the temperature on the reaction:

$$CH_4(g) + H_2O(g) \rightleftharpoons CO(g) + 3H_2(g) \qquad \Delta H^{\ominus} = +206 \text{ kJ mol}^{-1}$$

(i) It shifts the equilibrium to the right.

(ii) It speeds up the endothermic reaction and slows down the exothermic reaction.

(iii) It increases the rate constant for both reactions

(iv) It lowers the activation energy of the forward reaction.

Q10

In the reaction:

$$HCO_3^- + H_2PO_4^- \rightleftharpoons H_2CO_3 + HPO_4^{2-}$$

the species acting as a base is/are:

(i) $H_2PO_4^-$

(ii) HPO_4^{2-}

(iii) H_2CO_3

(iv) HCO_3^-

Q11

Which of the following is/are true about the group 7 elements chlorine, bromine and iodine?

(i) The ionisation energies decrease chlorine to iodine

(ii) HI is a stronger acid than HBr which is stronger than HCl

(iii) The radius of the anions increases chloride to iodide.

(iv) Iodide ions are weaker reducing agents than bromide ions which are weaker reducing agent than chloride ions.

The answers to all these Questions to try are on pages 93–96.

12 Answers to Questions to try

> **Notes:** A tick against part of an answer indicates where a mark would be awarded.

Chapter 1 Organic chemistry – reactions

Q1 How to score full marks

(a) There are several alternative answers depending upon which reaction you choose:

	I	II	III
Reagent	conc. HNO_3/H_2SO_4 ✔	Br_2 (l) ✔	C_2H_5Cl or CH_3COCl ✔
Equation	$\bigcirc$ + $HNO_3 \rightarrow$ $\bigcirc$ –NO_2 + H_2O ✔	$\bigcirc$ + $Br_2 \rightarrow$ $\bigcirc$ –Br + HBr ✔	$\bigcirc$ + $C_2H_5Cl \rightarrow$ $\bigcirc$ –C_2H_5 + HCl ✔ $\bigcirc$ + $CH_3COCl \rightarrow$ $\bigcirc$ –$COCH_3$ + HCl
Conditions	Warm to 55°C ✔	Fe, or $FeBr_3$ or $AlCl_3$ catalyst, dry ✔	anhydrous $AlCl_3$ ✔
Name of product	nitrobenzene ✔	bromobenzene ✔	ethylbenzene or phenylethanone ✔

(b) **(i)** NO_2^+ or Br^+ or $C_2H_5^+$ or CH_3C^+CO ✔

(ii) For the production of the NO_2^+ electrophile and the mechanism, see page xxx

bromination $2Fe + 3Br_2 \rightarrow 2FeBr_3$ ✔ Friedel–Crafts $AlCl_3 + C_2H_5Cl \rightarrow CH_3CH_2^+ + AlCl_4^-$ ✔

then $Br_2 + FeBr_3 \rightarrow Br^+ + FeBr_4^-$ ✔ or $AlCl_3 + CH_3COCl \rightarrow CH_3C^+O + AlCl_4^-$ ✔

(iii)

curly arrow from circle **inside** benzene ring to **correct** atom in electrophile; ✔
intermediate with incomplete circle and positive charge; ✔ arrow from σ bond with
H atom to inside the benzene ring (loss of H^+) ✔

(c) Wear gloves because of corrosive substances; do experiment in fume cupboard
(toxic fumes). ✔

(d) **For ethanone:** any reaction of a ketone, e.g. addition of HCN in the presence of base
(or KCN) ✔
$\bigcirc$ –$COCH_3$ + HCN → $\bigcirc$ –$CH(OH)CN$ ✔
For ethylbenzene: chlorination e.g. $\bigcirc$ –C_2H_5 + $Cl_2 \rightarrow$ $\bigcirc$ –C_2H_4 + Cl + HCl ✔
in the presence of UV light.

(e) **(i)** $3 \times (-120) = -360$ kJ mol^{-1} ✔

(ii)

‘cyclohexatriene’ above benzene above cyclohexane; explanation: benzene is at a lower
energy level of 155 kJ mol^{-1} compared with ‘cyclohexatriene’; because it is stabilised by
delocalised electrons/resonance. ✔

Q2 How to score full marks

(a)

Geometric (*cis/trans*) isomerism; the double (π) bond prevents rotation. ✔

(b) The acid is covalent and has a fairly large non-polar chain, so it is not soluble in water. ✔
When sodium hydroxide is added, it forms a salt which is ionic and so it dissolves in water. ✔

(c) (i) $CH_3CH=CHCOOH + C_2H_5OH \rightarrow CH_3CH=CHCOOC_2H_5 + H_2O$ ✔

(ii) it is a catalyst. ✔

(d) $CH_3CH=CHCH_2OH$ for forming the primary alcohol group CH_2OH ✔ and for leaving C=C not reduced. ✔

Chapter 2 Organic synthesis, analysis and spectroscopy

Q1 How to score full marks

The reaction $\mathbf{A} \rightarrow \mathbf{B}$ is reduction. ✔ The reaction $\mathbf{B} \rightarrow \mathbf{C}$ is dehydration (elimination of water), ✔ so $\mathbf{B}$ is probably an alcohol. ✔

$\mathbf{A}$ has a C=O group because its infra-red spectrum has a peak at 1715 cm⁻¹ ✔ and $\mathbf{B}$ has a C–OH group because it has an absorption band at 3350 cm⁻¹. ✔

$\mathbf{A}$ is either a ketone or an aldehyde, but the NMR spectrum shows that there are hydrogen atoms in only two different environments. ✔ Thus its formula must be $CH_3CH_2COCH_2CH_3$. The two CH_3 groups are next to a CH_2 group, and so will split into a triplet, whereas the two CH_2 groups are next to a CH_3 group and so will split into a quartet. The two peaks are in the ratio of 2:3. This is explained by the fact that there are four CH_2 hydrogen atoms and six CH_3 hydrogen atoms, so their ratio is 4:6 which is the same as 2:3. ✔

$\mathbf{B}$ is an alcohol as it has only one oxygen atom and contains an OH group (I–R spectrum) and is dehydrated by concentrated sulphuric acid, so the formula for $\mathbf{B}$ is $CH_3CH_2CH(OH)CH_2CH_3$. ✔

In the mass spectrum of $\mathbf{A}$ the molecular ion fragments:
$(CH_3CH_2COCH_2CH_3)^+ \rightarrow (CH_3CH_2CO)^+$ ✔ $+ CH_3CH_2$ ✔

$\mathbf{B}$ is dehydrated by concentrated sulphuric acid into an alkene $\mathbf{C}$, ✔ which is pent-2-ene, $CH_3CH=CHCH_2CH_3$. $\mathbf{C}$ shows geometric (*cis/trans*) isomerism ✔ produced in equal amounts.

When this is hydrated under acidic conditions, the H⁺ adds on either to the 2- or to the 3-position, forming the intermediates $CH_3CH_2CH^+CH_2CH_3$ ✔ and $CH_3CH^+CH_2CH_2CH_3$. Both of these are secondary carbocations and so stable, and so are produced in equal amounts. ✔

The second step in the hydration produces two alcohols, **B** and **D**.

D is $CH_3CH(OH)CH_2CH_2CH_3$, which is chiral but the racemic mixture is produced. ✓

B is not chiral as there are two C_2H_5 groups on the central carbon atom. ✓

Examiner's comments

- Try to deduce the functional group from the molecular formula and from its reactions.
- Compound A contains only one oxygen atom, and so cannot be an acid. It must be an aldehyde, ketone or an alcohol. It could also contain a C=C group, but it doesn't. $NaBH_4$ reduces a C=O but not a C=C group, so A must be an aldehyde or ketone.
- Analyse the spectral data

    Infrared: look for a band at about $1700\ cm^{-1}$ (C=O) and a broad band at about $3000\ cm^{-1}$ (OH).

    NMR: the number of peaks at different δ values tells you the number of different types of H atoms, and the splitting of these peaks the number of neighbouring H atoms.

    Mass: look for (M–15), which is caused by loss of CH_3 and for (M–29), which is caused by loss of C_2H_5.

- In this question possibilities for A are: pentan-3-one, pentan-2-one, pentanal and methylbutanal. All except pentan-3-one would give 3 NMR peaks, but there are only two peaks, so A must be pentan-3-one
- Hydration of the alkene **C** gives two structural isomers **B** and **D**, but **D** has optical isomers. Always look for this type of isomerism in problems such as this.

Q2 How to score full marks

(a) (i) % C in $C_4H_5 = (48/53) \times 100 = 90.57\%$ % H in $C_4H_5 = 100 - 90.57 = 9.43\%$

These values are consistent with the data. ✓

(ii) The mass of $C_4H_5 = 53$. $106/53 = 2$, ✓ so the molecular formula is C_8H_{10}. ✓

(b)

✓   ✓   ✓   ✓

(c) (i) **A:** the H atoms in C_6H_5 cause this peak. ✓ The chemical shift is 7.2, near the value given in the data. ✓ There is no splitting because all 5 protons are in the same environment ✓ and the peak height is 5, indicating that 5 of the 10 protons are in this group. ✓

(ii) **B:** the H atoms in CH_2 cause this peak. ✓ The chemical shift is 2.5, which is consistent with a CH_2 group between a benzene ring and an alkyl group. ✓ The band is split into a quartet because of the adjacent CH_3 group ✓ and the peak height is 2, indicating that 2 of the 10 protons are in this group.

(iii) **C:** the H atoms in CH_3 cause this peak. ✓ The chemical shift is around 1, which is consistent with the data for a CH_3 group in an alkyl chain. ✓ The band is split into a triplet because of the adjacent CH_2 group ✓ and the peak height is 3, indicating the 3 protons. ✓

(d) Thus the hydrocarbon **E** is ethylbenzene, $C_6H_5CH_2CH_3$. ✓

Examiner's comments

(a) Another way is to calculate the empirical formula by dividing the % of each element by its atomic mass.

(b) Two substituents on a benzene ring give three isomers. The fourth isomer has a C_2H_5 group instead of two CH_3 groups.

(c) Look at your structures in **(b)**. The peak heights are the easiest clues. Don't worry if the chemical shifts are slightly different from those in the data. You must not only identify the protons causing the peaks A, B and C, but also explain fully your reasons.

Chapter 3 Chemical equilibrium

Q1 How to score full marks

(a) (i) $\Delta H_{reaction} = \Delta H_{formation}$ of ethene $- \Delta H_{formation}$ of ethane = a positive value, and so the reaction is endothermic. Therefore the value of K_p will increase with an increase in temperature.

(ii) K_p is only altered by temperature, and so a change in pressure will not affect the value of K_p. ✓

(b)
	C_2H_6	$\rightleftharpoons$	C_2H_4	+	H_2
moles at start	1.0		0		0
moles at equilibrium	1.0 − 0.36 = 0.64		0.36		0.36 ✓

total moles = 1.36

partial pressure = mole fraction × total pressure ✓

$p(C_2H_6) = \dfrac{0.64}{1.36} \times 180 = 84.71$ kPa ✓ $\qquad p(C_2H_4) = p(H_2) = \dfrac{0.36}{1.36} \times 180 = 47.65$ kPa ✓

$K_p = \dfrac{p(C_2H_4) \times p(H_2)}{P(C_2H_6)} = \dfrac{47.65 \times 47.65}{84.71} = 26.8$ kPa ✓

Examiner's comments

(a) Pressure may alter the position of an equilibrium, but it never alters the value K.

(b) Don't forget that you must use equilibrium not initial values when calculating K. Don't forget to multiply the mole fraction by the total pressure to get the partial pressure. If you set your calculation out clearly (as shown), you will be less likely to make a mistake.

Q2 How to score full marks

(a) The reaction is exothermic, and so a high temperature would result in a low yield. ✓ However, a low temperature would mean that the reaction would be very slow, ✓ so a catalyst is used which allows the reaction to proceed at a fast rate at a moderate temperature. ✓ A high pressure will drive the equilibrium to the right, producing a greater yield, ✓ but a very high pressure would be expensive, ✓ and so a pressure of 70 atmospheres is the most economic.

(b) If the gaseous mixture is cooled, ✓ the water and the ethanol condense leaving the ethene gas, which is then recycled ✓ by being mixed with more ethene and passed through the catalyst.

(c)
	C_2H_4	+	H_2O	$\rightleftharpoons$	C_2H_5OH
moles at start	1.00		1.00		0
moles at equilibrium	1.00 − 0.900 = 0.100		0.100		0.900 ✓

$[C_2H_4]_{eq} = [H_2O]_{eq} = 0.100/3 = 0.0333$ mol dm^{-3} $\qquad [C_2H_5OH]_{eq} = 0.900/3 = 0.300$ mol dm^{-3}

$K_p = \dfrac{[C_2H_5OH]_{eq}}{[C_2H_4]_{eq} \times [H_2O]_{eq}}$ ✓ $\quad = \quad \dfrac{0.300}{0.0333 \times 0.0333} = 271$ mol^{-1} dm^3 ✓

Examiner's comments

(a) Good economics means maximising rate and yield, but without the expense of a very high pressure.

(c) Water appears in the expression for K_c as it is a gas and not the solvent.

Chapter 4 Acid–base equilibria

Q1 How to score full marks

(a) $Ba(OH)_2 \rightarrow Ba^{2+}(aq) + 2OH^-(aq)$

$[OH^-] = 2 \times 0.123 = 0.246$ mol dm^{-3}; ✓ pOH = 0.61; pH = 14 − pOH = 13.39 ✓

(b) (i)

Starting pH between 2.5 and 3.5; ✓ vertical range between 5.5/6.5 and 11.5/12.5; ✓ equivalence point 10.0 cm^3 of Ba(OH)$_2$; correct shape of graph. ✓

(ii) Cresol red ✓ because pH of 8 ±1 is within the vertical part of the graph. ✓

(c) Base CH_3CO_2H/conjugate acid $CH_3CO_2H_2^+$; ✓ acid H_2SO_4/conjugate base HSO_4^-. ✓

(d) When neutral $[H^+] = [OH^-]$ ✓

$[H^+] = \sqrt{(3.8 \times 10^{-14})}$ mol dm^{-3}; neutral pH at 40°C = 6.71 ✓

Examiner's comments

(a) Remember that barium hydroxide has two OH$^-$ ions per formula.

(b) When you draw these pH graphs, you must identify the starting pH, the range of the vertical part of the graph and the volume needed for equivalence (the end point volume).

(c) Acids give H$^+$; bases accept them.

(d) The pH of a neutral solution is 7 only at 25°C.

Q2 How to score full marks

(a) A strong acid is totally ionised in solution. ✓ Concentrated and dilute refer to the concentration of the solution, which is the number of moles in 1 dm^3 of solution. ✓ A concentrated solution has many more moles per dm^3 than a dilute solution. ✓

(b) (i) The oxidation number of phosphorus is +5. ✓

(ii) moles of $H_3PO_4 = 0.100 \times 20.0/1000 = 0.00200$ mol ✓

moles of NaOH = 3 × moles H_3PO_4 = 3 × 0.00200 = 0.00600 mol ✓

volume of NaOH = 0.00600/0.250 = 0.0240 dm^3 = 24.0 cm^3 ✓

(c) $[OH^-] = 0.250$ mol dm^{-3} then:

either pOH = −log(0.250) = 0.60; ✓ pH = 14 − pOH = 13.40 ✓

or $[H^+].[OH^-] = 1.00 \times 10^{-14}$; ✓ $[H^+] = 1.00 \times 10^{-14}/0.250 = 4.00 \times 10^{-14}$ ✓

pH = −log (4.00×10^{-14}) = 13.40 ✓

Examiner's comments

(a) You can have a concentrated solution of a weak acid – lots of moles but little ionisation – and a dilute solution of a strong acid – few moles and fully ionised.

(b) (i) Each of the three hydrogen atoms is +1 = +3, and each of the four oxygen atoms is −2 = −8. The oxidation numbers add up to zero, so phosphorus must be +5.

(ii) Remember that, for a solution, moles = concentration × volume in dm^3.

(c) Don't forget to show all your working. You won't get full marks if you don't. If you make a mistake, you can only get some marks if you have your calculation set out clearly and fully.

Q1 How to score full marks

(a) **(i)** From experiments 1 & 2: $[NO_2]$ is doubled while $[CO]$ is constant, and the rate is increased by a factor of 4, so the order with respect to NO_2 is two. ✓

From experiments 1 & 3: $[CO]$ is doubled while $[NO_2]$ is constant, and the rate is unaltered, so the order with respect to CO is zero. ✓

The rate equation is: rate of reaction = $k [NO_2]^2$ ✓

(ii) $k = \dfrac{\text{rate}}{[NO_2]^2} = \dfrac{1.3 \times 10^{-5} \text{ mol dm}^{-3} \text{ s}^{-1}}{(1.2 \times 10^{-2} \text{mol dm}^{-3})^2} = 0.090 \text{ mol}^{-1} \text{ dm}^3 \text{ s}^{-1}$ ✓

(b) If mechanism I is correct, the rate equation would be: rate = $k [NO_2] [CO]$, therefore mechanism I is wrong. ✓

If mechanism II is correct, the rate = the rate of the slowest step; so rate = $k [NO_2]^2$, which is supported by the rate equation obtained from the kinetic data. ✓

(c) **(i)** Step 1, because it is the slower step. ✓

(ii) Step 2, because it is the faster reaction. ✓

> **Examiner's comments**
>
> **(a) (i)** You must show how you obtained the order for each reactant.
>
> **(ii)** Be careful about the units for k. If you put units into the calculation, you should get them correct. The units of k depend upon the total order of the reaction.
>
> **(b)** You can get the order for a single step from the stoichiometry of that step.
>
> **(c)** Remember that large activation energy means a slow reaction.

Q2 How to score full marks

(a) **(i)** The rate equation is: rate = $k [X] . [Y]^2$ ✓ ✓

(ii) The overall order is 3. ✓

(iii) If the concentrations of both are doubled, the rate will increase by a factor of $2^3 = 8$. ✓

(b) **(i)** Comparing experiments 1 & 2: the overall order is 2 because doubling both $[A]$ and $[B]$ causes the rate of the reaction to quadruple. ✓

(ii) Comparing experiments 2 & 3: doubling $[A]$ while keeping $[B]$ constant, also quadruples the rate, so the reaction is second order with respect to A and hence zero order with respect to B. ✓

(iii) Rate = $k [A]^2$ ✓

(iv) From experiment 1: $k = \dfrac{\text{rate}}{[A]^2} = \dfrac{3.5 \times 10^{-4} \text{ mol dm}^{-3} \text{ s}^{-1}}{(0.2 \text{ mol dm}^{-3})^2} = 0.0088 \text{ mol}^{-1} \text{ dm}^3 \text{ s}^{-1}$ ✓

> **Examiner's comments**
>
> See comments on question 1 (a)

Q1 How to score full marks

(a)

$\Delta H_f = (\Delta H_a + 1^{st}\ IE)$ of Rb $+ (\Delta H_a + EA)$ of Cl + LE ✓

LE of RbCl $= (-435) - (+81) - (+403) - (+122) - (-349) = -692$ kJ mol⁻¹ ✓

All 6 steps correct and labelled = 4 marks; 5 or 4 steps = 3 marks; 3 steps correct = 2 marks; 2 steps only = 1 mark.

(b) The Li⁺ ion has a smaller radius than the Rb⁺ ion, and so the forces of attraction between the cation and the anion in the lattice are stronger in LiCl than in RbCl. ✓

> **Examiner's comments**
>
> **(a)** You must indicate what each step represents, either by labelling ΔH_a etc. or by putting in the actual values. Show all your working. Be very careful with the signs. Remember that lattice enthalpies are defined for the process ions → solid, and so are negative (exothermic).
>
> **(b)** You must not only say that the forces are stronger in LiCl, but also why.

Q2 How to score full marks

(a) At 100°C, $\Delta G = 0$ for the reaction $H_2O(g) \rightleftharpoons H_2O(l)$ therefore $\Delta H = T\Delta S$. ✓

$\Delta S = S_{products} - S_{reactants} = S_{water} - S_{steam} = 70 - 189 = -119$ J K⁻¹ mol⁻¹. ✓

Enthalpy of condensation $= \Delta H = T\Delta S = 373 \times (-119) = -44.4$ kJ mol⁻¹. ✓

Below 100°C the term $(-T\Delta S)$ becomes smaller, but ΔH is still -44.4 kJ mol⁻¹.

Because $\Delta G = \Delta H - T\Delta S$, ΔG becomes negative and the process is spontaneous. ✓

(b) The reaction is $CH_4(g) + H_2O(g) \rightarrow CO(g) + 3H_2(g)$ ✓

The reaction becomes spontaneous when ΔG changes from + to − ✓

$\Delta S = (198 + 3 \times 131) - (186 + 189) = +216$ J K⁻¹ mol⁻¹ ✓

since $\Delta G = \Delta H - T\Delta S$ and ΔH is positive ΔG will only become negative when $T\Delta S$ becomes $> \Delta H$, which is at high temperatures. ✓

(c) The reaction $C_{diamond} \rightarrow C_{graphite}$ has $\Delta S = 6 - 3 = +3$ J K⁻¹ mol⁻¹ ✓

As ΔH is negative and as $\Delta G = \Delta H - T\Delta S$, ΔG will always be negative and so the reaction will always be thermodynamically feasible. However, the reaction has a very large activation energy, and so is so slow that it is not observed at room temperature. ✓

(d) The reaction is $CaO(s) + CO_2(g) \rightarrow CaCO_3(s)$ and $\Delta S = 90 - (40 + 214) = -164$ J K⁻¹ mol⁻¹

When $\Delta G = 0$, $\Delta H = T_s\Delta S$ therefore $T_s = \Delta H/\Delta S = (-178 \times 10^3)/(-164) = 1085$ K ✓

> **Examiner's comments**
>
> For all four parts you must write balanced equations, and you must show all your working.
>
> For all, you must know that a reaction becomes spontaneous at the point when $\Delta G = 0$, which is when $\Delta H = T\Delta S$.
>
> **(b)** If ΔH and ΔS are both positive, the reaction will become spontaneous at a high enough temperature to make the $T\Delta S$ term outweigh the ΔH term.
>
> **(d)** If both are negative, the reaction will cease to be spontaneous if the temperature gets too high.

Chapter 7 The periodic table – period 3

Q1 How to score full marks

(a) (i) Equation: $Na_2O + H_2O \rightarrow 2\,NaOH$ (or $Na_2O + H_2O \rightarrow 2Na^+ + 2OH^-$) ✓

pH = 13 or 14 ✓

(ii) Equation: $SO_2 + H_2O \rightarrow H_2SO_3$ ✓

pH = 3 ✓

(b) Ionic oxides, such as sodium oxide, react to form alkaline solutions ✓ with pH > 7.

Covalent oxides, such as sulphur dioxide, react to form acids ✓ with pH < 7.

(c) (i) Equation: $MgCl_2 + aq \rightarrow Mg^{2+}(aq) + 2Cl^-(aq)$ ✓

pH = 7 ✓

(ii) Equation: $SiCl_4 + 4H_2O \rightarrow Si(OH)_4 + 4HCl$

or $SiCl_4 + 2H_2O \rightarrow SiO_2 + 4HCl$ ✓ (or $+ 4H^+(aq) + 4Cl^-(aq)$ on right-hand side)

pH = 1 or 2 ✓

Examiner's comments

You may write full or ionic equations in this question, as neither was specified. State symbols do not have to be added, but if you do add them they must be correct.

(a) NaOH is a strong base and is totally ionised; H_2SO_3 is a weak acid and is partially ionised, and so its pH is more than that of a strong acid.

(b) and (c) Oxides and chlorides of the period 3 elements behave differently with water according to their bonding. $MgCl_2$ is ionic and just dissolves, producing neither H^+ nor OH^- ions and so its solution is neutral.

Chapter 8 Redox equilibria

Q1 How to score full marks

(a) (i) $Fe^{3+}(aq) + e^- \rightarrow Fe^{2+}(aq)$; ✓
$MnO_4^-(aq) + 8H^+(aq) + 5e^- \rightarrow Mn^{2+}(aq) + 4H_2O(l)$ ✓

(ii) $MnO_4^-(aq) + 8H^+(aq) + 5Fe^{2+}(aq) \rightarrow MnO_4^-(aq) + 4H_2O(l)$ species; ✓ balance ✓

(b) Moles of $KMnO_4 = 0.0655 \times 23.4/1000 = 0.0015327$ ✓

Moles of $Fe^{2+} = 5 \times 0.0015327 = 0.007664$ in $25\,cm^3$ of solution ✓

Moles of Fe^{2+} in $250\,cm^3$ of solution = 0.07664 = moles of iron in sample ✓

Mass of iron in sample = $0.07664\,mol \times 55.8\,g\,mol^{-1} = 4.277\,g$ ✓

Purity = $(4.277/4.45) \times 100 = 96.1\%$ ✓

Examiner's comments

(a) Half-equations should be written as reduction reactions with electrons on the left. The oxidation number of iron changes by 1, so there is 1 electron in its half equation. The oxidation number of manganese changes from +7 to +2 (by 5), and so there are 5 electrons in its half equation.

To get the overall equation, the first half-equation must be reversed and multiplied by 5 and then added to the second half-equation. Remember that the reactants are MnO_4^- and Fe^{2+} (not Fe^{3+}) and so go on the left of the overall equation. You would lose 1 mark if state symbols were not present in your equations.

(b) The ratio of Fe^{2+} to MnO_4^- is 5:1, and so the moles of MnO_4^- must be multiplied by 5 to get moles of Fe^{2+}. Don't forget that the solution was diluted, and so you have to multiply by 10 to get the total moles.

Chapter 9 Transition metals

Q1 How to score full marks

(a) (i) The electronic structures are: Cu [Ar] $3d^{10} 4s^1$, and for Cu^+ [Ar] $3d^{10}$ ✓

(ii) The copper atom has ten 3d electrons and only one 4s electron, whereas all the other d-block elements (except Cr) have two 4s electrons. ✓

(iii) As the 3d level is full, it is not possible for an electron to be promoted from one d orbital to another, and so the ion will not absorb a photon in the visible region. ✓

(b) E°_{cell} for the reaction $2Cu^+(aq) \rightarrow Cu(s) + Cu^{2+}(aq)$ is the sum of the two half equations

$$Cu^+(aq) + e^- \rightarrow Cu(s) \qquad E^\circ = + 0.52V$$
$$Cu^+(aq) \rightarrow Cu^{2+}(aq) + e^- \qquad E^\circ = - (+0.15V) = -0.15V \quad \left. \right\} ✓$$

$E^\circ_{cell} = =0.52 + (- 0.15) = + 0.37$ V, which is positive and so the reaction will take place. ✓

Thus the Cu^+ ions in the Cu_2SO_4 will disproportionate in aqueous solution. And you would observe the colourless Cu_2SO_4 producing a pink/brown precipitate in a blue solution. ✓

(c) (i)

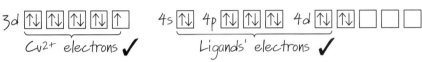

Cu2+ electrons ✓   Ligands' electrons ✓

(ii) First a pale blue precipitate will be produced by the reaction $[Cu(H_2O)_6]^{2+} + 2NH_3 \rightarrow Cu(OH)_2 + 2NH_4^+ + 4H_2O$, which is a deprotonation reaction. Then, with excess ammonia, a deep blue solution is formed by the reaction $Cu(OH)_2 + 4NH_3 + 2H_2O \rightarrow [Cu(NH_3)_4(H_2O)_2]^{2+} + 2OH^-$, which is ligand exchange. ✓

Examiner's comments

(a) (i) There is no need to write out all the core 1s, 2s, 2p, 3s and 3p electrons unless a full electronic configuration is asked for.

(iii) Remember that the colour is caused by a d-d transition, and that this is not possible here as there is no space in the upper d level as the d shell is completely full.

(b) The value of E_{cell} can be calculated using the 'anticlockwise' rule, but you must still write out the two half equations, and in the correct order.

(c) (i) Make it clear which are the copper electrons and which are the ligands'.

(ii) The last step is called ligand exchange, because the overall reaction is that the H_2O ligands have been replaced by NH_3 ligands.

Q2 How to score full marks

(a) You can illustrate your answer with any transition element. Below are answers for copper and iron.

	Copper	Iron
Complex ion formation ✓	$[Cu(NH_3)_4(H_2O)_2]^{2+}$ or $[CuCl_4]^{2-}$	$[Fe(CN)_6]^{3-}$ or $[Fe(CN)_6]^{4-}$ or $[Fe(SCN)(H_2O)_5]^{2-}$
Variable oxidation state ✓	Cu^{2+}, Cu^+ ✓	Fe^{3+}, Fe^{2+} (and FeO_4^-)
Colour of ions ✓	$[Cu(H_2O)_6]^{2+}$ is blue ✓	$[Fe(H_2O)_6]^{3+}$ is amethyst; $[Fe(H_2O)_6]^{2+}$ is pale green
Catalytic activity ✓	Copper metal is a catalyst for the dehydrogenation of alcohols ✓	Iron is the catalyst for the Haber process (the manufacture of ammonia)

(b) **A** is hydrated copper(II) chloride, $CuCl_2.2H_2O$ ✓

B is $[Cu(H_2O)_6]^{2+}$ ions, ✓   and **Z** is $[Cu(NH_3)_4(H_2O)_2]^{2+}$ ions ✓

X is silver chloride, AgCl ✓   and **Y** is copper(II) hydroxide, $Cu(OH)_2$ ✓

Examiner's comments

(a) Vanadium would be another example, although it does not form many complex ions and you would have to give an aqua complex as your example.

(b) In this question either formulae or names are acceptable. Copper chloride contains a mixture of $[Cu(H_2O)_4]^{2+}$ ions, which are blue, and $[CuCl_4]^{2-}$, which are yellow, making the solid and concentrated solution green. When diluted, ligand exchange takes place producing the aqua ion. The precipitate **X** is formed in the standard chloride ion test.

Chapter 11 Synoptic questions

Q1 How to score full marks

Bonding: Ionic compounds have a high boiling point because the force between the ions is a very strong force. ✓ Sodium chloride has a giant ionic lattice, ✓ ethanoic acid and ethanol are covalent organic molecules and aluminium chloride is also covalent ✓ (because of the fairly small difference in electronegativities of Al and Cl). All three form simple molecular lattices ✓ with hydrogen bonds between molecules in ethanol and ethanoic acid ✓ and London dispersion and dipole/dipole forces in aluminium chloride, ✓ and so all three have boiling points lower than sodium chloride.

Solutions:

Sodium chloride dissolves in water producing separate ions: $NaCl(s) \rightarrow Na^+(aq) + Cl^-(aq)$, ✓ which can therefore move, and so conduct electricity. ✓ It is the salt of a strong acid and a strong base, and so the solution is neutral, $[H^+] = 1 \times 10^{-7}$ mol dm^{-3} or pH = 7. ✓

Ethanoic acid is a weak acid and so partially ionises $CH_3COOH \rightleftharpoons H^+ + CH_3COO^-$, ✓ producing enough ions to cause it to conduct, but not enough to be a good conductor. ✓ The solution has a pH = $-\log (1.3 \times 10^{-3}) = 2.88$. ✓

K_a for ethanoic acid = $[H^+] \times [CH_3COO^-]/[CH_3COOH]$, ✓

and as $[H^+] = [CH_3COO^-]$, $K_a = (1.3 \times 10^{-3})^2/0.1 = 1.7 \times 10^{-5}$ mol dm^{-3}. ✓

Ethanol does not form ions in water, and so is not a conductor of electricity. ✓ It does not react with water, and so the pH of its solution is the same as that of pure water, which is 7. ✓

Aluminium chloride reacts rapidly with water $AlCl_3 + 3H_2O \rightarrow Al(OH)_3 + 3H^+ + 3Cl^-$. ✓ As ions are formed the solution will conduct electricity well. ✓ As H^+ ions are formed, the solution will be acidic, ✓ and its pH = $-\log (0.30) = 0.52$. ✓

Examiner's comments

• To a maximum of 19 (but something must be said about all four substances in each heading).

• Quality of language scores an extra 3 marks, making a total score of 22.

• The answer explained the difference between the high and lower boiling points in terms of the bonding and hence the forces between particles, and the conductivity in terms of the number of ions produced on dissolving.

• The $[H^+]$ data was explained in terms of whether they react with water to form an acid, and it was also used quantitatively as the pH of each solution was calculated.

• The answer was written in continuous prose, properly punctuated, and the chemical principles were clearly explained. Thus three 'quality of language' marks would have been awarded.

Q2 How to score full marks

(a) (i) The long alkyl chain will tend to make bombykol insoluble in water, ✓ but the OH group will hydrogen bond with water causing solubility. Thus bombykol will be slightly soluble. ✓

(ii) It will form 4 ✓ geometric isomers. Each of the double bonds will cause a *cis/trans* pair. ✓

(iii) *Alkenes*, such as bombykol, react with:

hydrogen ✓ in the presence of a platinum (or nickel) catalyst:

$$RCH=CH-CH=CHR' + 2H_2 \rightarrow RCH_2-CH_2-CH_2-CH_2R' \checkmark$$

(where R stands for $CH_3(CH_2)_2$ and R' for $(CH_2)_8CH_2OH$)

or hydrogen bromide: ✓

$$RCH=CH-CH=CHR' + 2HBr \rightarrow RCHBr-CH_2-CH_2-CHBrR' \checkmark$$

or bromine: ✓

$$RCH=CH-CH=CHR' + 2Br_2 \rightarrow RCHBr-CHBr-CHBr-CHBrR' \checkmark$$

Alcohols, such as bombykol, react with:

acidified potassium dichromate, when heated under reflux: ✓

$$R''CH_2OH + 2[O] \rightarrow R''COOH + H_2O \checkmark$$

or sodium: ✓

$$R''CH_2OH + Na \rightarrow R''CH_2O^-Na^+ + \tfrac{1}{2}H_2 \checkmark$$

or phosphorus pentachloride: ✓

$$R''CH_2OH + PCl_5 \rightarrow R''CH_2Cl + POCl_3 + HCl \checkmark$$

or ethanoic acid in the presence of concentrated sulphuric acid: ✓

$$R''CH_2OH + CH_3COOH \rightarrow CH_3COOCH_2R'' + H_2O \checkmark$$

To a maximum of 4 marks as long as you give at least one reaction for each functional group.

(b) The part of the molecule to the left of the first double bond will give butanoic acid, $CH_3(CH_2)_2COOH$. The part between the two double bonds will produce ethandioic acid, HOOCCOOH. The part to the right of the second double bond will produce decandioic acid, $HOOC(CH_2)_8COOH$.

Names or formulae of all three acids = 2 marks; of any two acids = 1 mark.

Examiner's comments

(a) (i) You are not expected to know its solubility, but to deduce the effect of the OH group and the long carbon chain on solubility.

(ii) Each double bond causes two isomers, so there are four possible geometric isomers.

(iii) Other reactions of alkenes and alcohols are acceptable, but you must remember that both double bonds will react.

(b) The double bonds break, initially producing a CHO group where there was a C=. This gives two molecules each with one CHO group and a molecule with a CHO group on both ends. Then these, and the original OH group, become oxidised to COOH groups.

Q3 How to score full marks

(a)

	% by mass	÷ atomic mass = moles	÷ smallest
C	49.3	$49.3 \div 12 = 4.11$ ✔	3.0
N	19.2	$19.2 \div 14 = 1.37$	1.0 Empirical formula is C_3H_7NO ✔
H	9.6	$9.6 \div 1 = 9.6$	7.0
O	21.9	$21.9 \div 16 = 1.37$	1.0

(b) The equivalence (end) point is the volume corresponding to half-way up the vertical part of the graph. Here it is $25.0\,cm^3$. ✔
Moles of HCl = $0.104\,mol\,dm^{-3} \times 0.0200\,dm^3 = 0.00208$. ✔
A primary amine reacts with HCl in a 1:1 ratio, therefore moles of **X** = 0.00208.
[**X**] = $0.00208\,mol/0.0250\,dm^3 = 0.0832\,mol\,dm^{-3}$. ✔
Molar mass of **X** = $6.07/0.0832 = 73.0\,g\,mol^{-1}$. ✔

(c) **(i)** mass of C_3H_7NO is also 73, therefore the molecular formula is C_3H_7NO. ✔

(ii) It does not contain a C=C so two possibilities for **X** are:
$$CHOCH_2CH_2NH_2 \text{ ✔ and } CH_3COCH_2NH_2 \text{ ✔}$$
         I   **II**

(d) **(i)** The C=O group can be identified by adding 2,4-dinitrophenylhydrazine solution. ✔
A red/orange precipitate would be obtained. Adding Fehling's solution would distinguish the two isomers for **X**. The aldehyde, $CHOCH_2CH_2NH_2$, would give a red-brown precipitate of copper(I) oxide on warming, ✔ whereas the ketone, $CH_3COCH_2NH_2$ would not. ✔

(ii) Both would give an ion of m/e of 57 by loss of NH_2. ✔ However, only one of the isomers will give the peak at m/e 58, which is 15 less than the molecular ion and is caused by loss of CH_3 forming a $COCH_2NH_2^+$ ion. ✔ Only **II** has a CH_3 group and so **X** is structure **II**. ✔

Examiner's comments

(b) The end point is half way up the vertical part of the graph. Here the salt of a weak base and a strong acid is produced, and this has a pH < 7. The end point is only at pH = 7 for a strong acid/strong base titration.
The route for the calculation is: moles of HCl → moles of X → concentration of X = moles in 1 dm^3 → molar mass using mass of X in 1 dm^3

(c) (ii) The substance cannot be $CH_3CH_2CONH_2$ as this is an amide not an amine.

(d) (i) First you must prove the presence of the C=O group and then you must test to distinguish an aldehydes from a ketone

(ii) You must see what has been broken off from the molecular ion. 15 is CH_3 and 16 is NH_2.

Q4 How to score full marks

(a) moles of $NH_4Cl = 0.100\,mol\,dm^{-3} \times 0.100\,dm^3 = 0.0100\,mol$ ✔

moles of $NaNO_2 = 0.111\,mol\,dm^{-3} \times 0.080\,dm^3 = 0.00888\,mol$ ✔

As they react in a 1:1 ratio, the $NaNO_2$ is the limiting reagent as there is less of it. ✔

moles of N_2 produced = moles of $NaNO_2 = 0.00888\,mol$ ✔

volume of $N_2(g)$ at room T and P = $0.00888\,mol \times 24\,dm^3mol^{-1} = 0.213\,dm^3$ or $213\,cm^3$ ✔

(b) **Apparatus:**

1 mark for a suitable flask; 1 mark for a gas syringe or graduated tube; 1 mark for an apparatus that works (diagram 3 marks).

Method:

1 Pipette 50 cm^3 of one solution into the flask. ✔

2 Pipette 50 cm^3 of the other solution (into a beaker) and then add it to the solution in the flask. ✔

3 Insert the delivery tube in the flask, start a stop watch and collect the gas produced. ✔

4 Measure the volume of gas every minute. ✔

5 Plot a graph of volume of nitrogen against time. ✔

6 Measure the slope of the graph. This equals the rate of the reaction. ✔

7 Repeat, but first warm **both** solutions to 35°C in a thermostat. ✔

8 Plot its graph. The ratio of the slopes is the ratio of the rates at the two temperatures. ✔

Description to a maximum of 7 marks.

Examiner's comments

(a) You first have to work out which is the limiting reagent. If you had based your calculation on the moles of NH_4Cl, you would have scored only 3 marks for getting the answer 0.240 dm^3 (240 cm^3).

(b) The question asked you to select a suitable scale of apparatus and amount of the two solutions. Gas syringes and graduated tubes normally have a volume of 100 cm^3, but if you had chosen 50 cm^3 apparatus and hence 25 cm^3 of each solution, you would have scored full marks.

You must give full details of your method, and you must say how you can use the results to find out how much faster the reaction is at the higher temperature. It is a good idea to number or bullet the various operations in the method.

Q5–11 How to score full marks

5 C. 6 D. 7 A 8 B. 9 B. 10 C. 11 A.

Examiner's comments

5 $KBiO_3$ is reduced, the Bi atom going from an oxidation number of +5 to +3. Oxidising agents get reduced, so the answer is C.

6 In $(CH_3)_3N$, none of the H atoms is bonded to an F, O or N atom.

7 Solutions must be 1 mol dm^{-3} to get the standard cell potential.

8 A negative value of ΔS means that the entropy (disorder) has decreased.

9 An increase in temperature speeds up both the exo- and endothermic reactions. Only a catalyst will lower the activation energy.

10 The reaction is reversible, so you must look for a base on **both** sides of the equation. Bases accept H$^+$ ions.

11 The strength of the acids in group 7 increase down the group as the H-halogen bond gets weaker.